AF556010

FUNGI CLASSIFICATION AND IDENTIFICATION

a division of
NIPA GENX ELECTRONIC RESOURCES & SOLUTIONS P. LTD.
New Delhi-110 034

OM SHANTI

FUNGI CLASSIFICATION AND IDENTIFICATION

Reeti Singh
Ajay Kumar
Jagdish Kumar Patidar
Pragati Saini
R.K. Pandya
Ashish Bobade
Radha Gupta

Department of Plant Pathology
College of Agriculture
Rajmata Vijayaraje Scindia Krishi Vishwa Vidyalaya
Gwalior, Madhya Pradesh, India

a division of
NIPA GENX ELECTRONIC RESOURCES & SOLUTIONS P. LTD.
New Delhi-110 034

a division of

NIPA GENX ELECTRONIC RESOURCES & SOLUTIONS P. LTD.

101,103, Vikas Surya Plaza, CU Block
L.S.C.Market, Pitam Pura, New Delhi-110 034
Ph : +91 11 27341616, 27341717, 27341718
E-mail:newindiapublishingagency@gmail.com
www: www.nipabooks.com

For customer assistance, please contact
Phone: + 91-11-27 34 17 17 Fax: + 91-11- 27 34 16 16
E-Mail: feedbacks@nipabooks.com

ISBN: 978-93-89130-47-8

Composed and Designed by NIPA.

राजमाता विजयाराजे सिंधिया कृषि विश्वविद्यालय
राजा पंचम सिंह मार्ग, ग्वालियर (म.प्र.) – 474002
Rajmata Vijayaraje Scindia Krishi Vishwa Vidyalaya,
Raja Pancham Singh Marg, Gwalior (M.P.) – 474 002
(An ISO Certified 9001:2008)
Tel: 0751 – 2970502, Fax: 0751 – 2970504, E-mail: vcrvsaugwa@mp.gov.in

प्रो. एस. के. राव
कुलपति
Prof. S. K. Rao
Vice-Chancellor

No./VC/2018-19/ 581
Date: 01/07/2019

Foreword

The book entitled authored by Dr. Reeti Singh, Dr. Ajay Kumar, Dr. Jagdish Kumar Patidar, Dr. Pragati Saini, Dr. R.K. Pandya, Dr. Ashish Bobade and Dr. Radha Gupta is a timely publication to cater the need of students and research workers. The language of the book is simple, lucid and presents an appropriate expression of the subject. It gives me immense pleasure to introduce the book entitled **"Fungi: Classification & Identification"** authored by the team of the Department of Plant Pathology, College of Agriculture, Rajmata Vijayaraje Scindia Krishi Vishwa Vidyalaya, Gwalior, M.P., India. The book will benefit the students who wish to get knowledge of general classification of fungi and their descriptions, nutrition, reproduction, life cycle, culture media, electron microscopy and preservation of culture of fungi.

The chapters have been suitably illustrated with both black and white photographs along with line diagrams. Each chapter begins with fairly detailed, well organized diagrams that can be used by students and instructor as an outline of the chapter. Some new information and changes have also been incorporated in this book. This book will be more useful to undergraduate, post graduate students and teachers of Plant Pathology.

-sd-
(S.K. Rao)

Preface

The literature on fungi is huge and expanding rapidly. Many undergraduate students do not have sufficient time to read original publications, and rely solely on the course teacher due to non-availability of illustrated practical book in this field. We have therefore, tried to give important and useful information at one place. The main emphasis of this book is to present the fungi in such a simple way which can be understood by students. This book is designed to fulfil the syllabus by covering various aspects of fungi. It provides something for everyone from beginners to advanced students and researchers. There has been a continuous need to add at least some additional recent text and more illustrations to the book for the subject which will benefit the students. Fortunately through the use of computer tremendous advances have been made particularly in the reproducibility of the photographs and diagrams for the effective teaching. Mycology came alive when the diagrams of wide variety of fungi have been given at every possible place.

For this book, we drew heavily on numerous sources of information, Journals and the internet for illustrations. The taxonomic framework of the fungi has been based on the book, "The fungi: An Advance Treatise" vol. IV A and IV B written by G.C. Ainsworth, F.K. Sparrow and A.S. Sussman.

We are thankful to Rajmata Vijayaraje Scindia Krishi Vishwa Vidyalaya, Gwalior, M.P., India for giving the permission to write this book. We are grateful to our family members for their patience, involvement and untiring support.

The authors have attempted to trace and acknowledge the copy right holder of all material reproduced in this publication and apologize to copy holders if permission and acknowledgement to publish in this form have not been obtained. If any copyright material has not been acknowledged, please write and let us know so that we may rectify it.

There may be many errors and omissions and the authors will be highly indebted to those who will put their healthy suggestions in the improvement of the book. Such types of criticism/suggestions/ideas are always welcome by the authors.

Our sincere thanks to Shri Sarman Karosiya, Computer Operator, College of Agriculture, Gwalior for desirable typesetting of the book.

Authors

Contents

1

Fungi

1.1 Introduction

Fungi are chlorophyll-less thallophytic plant. Due to absence of chlorophyll, they are heterophytes which depend on other form of life for food. They grow in various habitats and show much diversity in their structure, physiology and reproduction. They evoloved long back from the ancient time.

Their existence was recorded from Pre-cambrian period. Information from ancient literature indicates that the fungi were used as food by men. At present, the fungi are used in medicine as well as food in addition to other aspects. The fungi cause diseases in crop plants (spots, rusts, smuts etc.) and on human beings (Aspergillosis, Blastomycosis etc.).

At present biologists use the term fungus (fungus mushroom, from Greek, sponagos = sponge) to include eukaryotic, spore bearing, achlorophyllous organism that generally reproduce sexually and asexually and whose usually filamentous, branched somatic structure are typically surrounded by cell wall containing chitin or cellulose or both of these substances, togather with many other complex molecules

More than 1.5 million species (Hawksworth 2001; Kirk *et al.*, 2001) of fungi have been recorded, but their number may be much more than the actual record. The subject which deals with fungi is known as Mycology (mykes — mushroom; logos—study) and the concerned scientist is called mycologist. The various definitions of fungi as proposed by mycologists are:

Alexopoulos (1962) defined fungi as "nucleated, spore-bearing, achlorophyllous organisms which generally reproduce sexually and asexually and whose usually filamentous, branched somatic structures are typically surrounded by cell walls containing cellulose or chitin or both".

Alexopoulos and Mims (1979) defined fungi as eukaryotic spore bearing, achlorophyllous organisms that generally reproduce sexually and asexually, and whose usually filamentous, branched somatic structures are typically surrounded by cell walls containing chitin or cellulose, or both of these substances, together with many other complex organic molecules.

1.2 General characters of fungi

The nutrition of fungi are heterotrophic absorptive. Vegetative state on or in substrate as a non motile mycelium except in chitrids, yeast, Plasmodiophoromycetes and slime mold. Motile reproductive units may occur. Cell wall is usually based on glucans and cellulose (Oomycetes). Nuclear status eukaryotic, uni- or multinucleate, the thallus being homo -or hetero- karyotic, haploid, dikaryotic or diploid. Life cycle is simple, usually complex. They reproduce sexually, asexually and parasexually. Propagules are typically microscopically small spores produced in high numbers. Motile spores are confined to certain groups. Sporocarp microscopic or macroscopic and showing characteristic shapes but only limited tissue differentiation. Habitat ubiquitous (seen everywhere) in terrestrial and freshwater habitat, less so in the marine environment. Important ecological roles as saprotrophs, mutualistic symbionts, parasites or hyper parasites. Distribution cosmopolitan in world.

1.3 Somatic structure of fungi

Vegetative body of fungi is known as thallus or soma. (thallus Gr.= shoot soma Gr. = body). Thallus is a simple plant body devoid of chlorophyll and differentiation into root, stem, leaves etc. They lack vascular system, but the reproductive and vegetative structures can be differentiated from each other. Fungal thalli may be plasmodium, unicellular or multicellular.

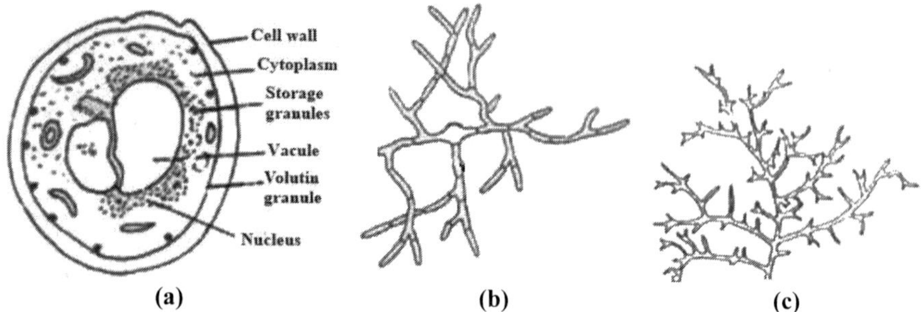

Fig.1.1. Different kind of thalli in fungi **(a)** Unicellular thallus of yeast **(b)** Septate branched hyphae **(c)** Mycelium with coenocytic branched hyphae

1.3.1 Plasmodium

Plasmodia vary in size and can be grouped into three categories.

1.3.1.1 Protoplasmodia

The plasmodium in some Myxomycetes (order Echinosteliales) is simple and of a primitive type. It is a uninucleate tiny mass of homogeneous slimy protoplasm which forms pseudopode but shows no distinction into veins. It is the smallest Myxomycetes and remains microscopic as long as it exists.

The cytoplasmic stream is indistinct, slow and irregular. At fruiting it gets converted into a single sporangium.

1.3.1.2 Aphanoplasmodia

This type of plasmodium is characteristic of the order Stemonitales. In early development the aphanoplasmodium looks very much like a protoplasmodium. During further growth it elongates and branches, finally resulting in an open network of delicate strands. The aphanoplasmodium is transparent, with individual strands only 5–10 μm wide and the entire plasmodiam is about 100–200 μm in diameter. The aphanoplasmodium lacks the slimy sheath. The plasmodial protoplasm is granular and thus transparent and not easily visible. The cytoplasmic streaming is rapid and confined by a fine membrane.

1.3.1.3 Phaneroplasmodia

It is the most common type and characteristic of the order Physarales. The mature phaneroplasmodium is a massive structure. In initial stages it is very much like the protoplasmodium. The protoplasm of phaneroplasmodium is multinucleate and highly granular.

It is differentiated into ectoplasm and endoplasm. At maturity it is divisible into an anterior fan shaped perforated sheet of protoplasm and posterior zone consisting of a reticulate network of tubular veins or strands which shows reverse cytoplasmic flow.

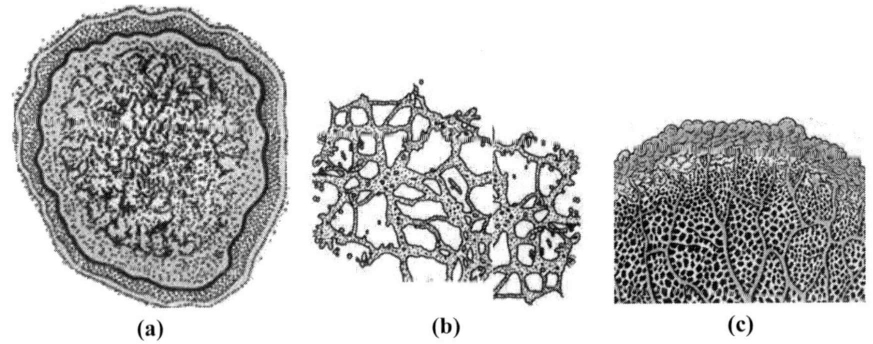

Fig.1.2. Types of plasmodia in Myxomycetes **(a)** Protoplasmodium **(b)** Aphanoplasmodium **(c)** Phaneroplasmodium

1.3.2 Unicellular

Thallus shows a single cell which is of various shapes and colour. These cells divide by fission or by budding ex. Yeast. A few species including certain pathogens of human and animals are dimorphic, that is capable of switching between hyphal and yeast like growth forms.

In *Candida parapsilosis* intermediate stage between yeast cells and true hyphae also occur and are termed as pseudohyphae. In lower fungi (Chitridiomycetes) grow as a thallus *i.e.*, a walled structure in which the protoplasm is concentrated in one or more centers, from which root like branches (rhizoids) ramify.

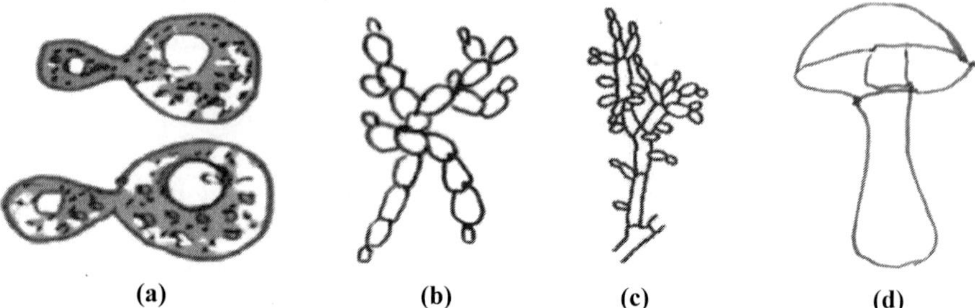

(a) (b) (c) (d)

Fig. 1.3. Different forms of mycelium **(a)** Unicellular Yeast **(b)** Pseudomycelium **(c)** Pseudohypha (*Candida*) **(d)** Mushroom

1.3.3 Multicellular

Majority of the fungi show multicellular or filamentous thalli. Thallus consists of microscopic threads or filaments which branch in all direction and spreading over or within the substratum utilized for food. Each of these filament is known as hypha (Hypha Gr. = web) *ex*. Mushroom.

1.3.3.1 Ectophyte

When fungus (thallus) is present on the host surface, it is known as ectophyte (on plant) or ectobiotic (on life). In case of ectobiotic mycelium, hyphae produce haustoria in epidermal cells.

1.3.3.2 Endophyte

If the hypha is present inside the host, it is endophyte or endobiotic. In endophytic mycelium, the hyphae may be inter – or intra-cellular. In intercellular condition the hyphae spread in between the cells while in intracellular condition the hyphae penetrate into the cells. Intercellular mycelium produces specialized organ, haustoria for absorption. They are absent in intracellular mycelium.

The mycelium may be colored or colorless (hyaline). Color of the hyphae is due to presence of pigments in cell wall or cell lumen.

1.3.3.3 Hypha

Hyphae are generally quite uniform in different taxonomic groups of fungi. They may or may not have cross walls or septa. The Oomycetes and Zygomycetes generally have aseptate, hyphae in which the nuclei lie in a common mass of cytoplasm. It is made up of a thin transparent, tubular filament which is filled or lined with protoplasm. Hypha is the unit of filamentous fungus and it grows only by apical elongation.

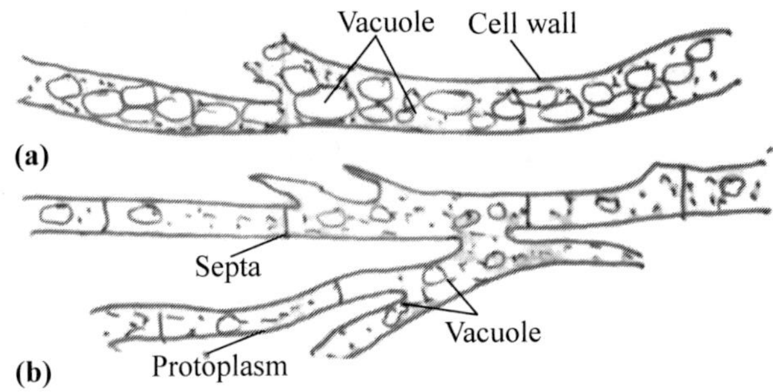

Fig. 1.4. Types of mycelium **(a)** Aseptate mycelium **(b)** Septate mycelium

1.3.3.3.1 Hyphal Branching

Assimilative hyphae of most fungi grow monopodially by a main axis (leading hypha) capable of potentially unlimited apical growth. Branches arise at some distance behind the apex. Dichotomous branching is rare but does occur in *Allomyces* and *Galactomyces.* In septate fungi branches are often located immediately behind a septum. Branches usually arise singly in vegetative hyphae, although whorls of branches (*i.e.*, branches arising near a common point) occur in reproductive structures. The circular appearance of fungal colonies in Petri dish cultures arises because certain lateral branches grow out and fill the space between the leading radial branches, keeping pace with their rate of growth. This invasive growth is the most efficient way to spread throughout a substratum.

Fungi can produce complex and characteristic multicellular structures which resemble the tissues of other eukaryotes. This must be controlled by the positioning, growth rate and growth direction of individual hyphal branches. It may be speculated that the diffusion of signaling molecules take place between adjacent hyphae. This may be facilitated by an extrahyphal glucan, matrix within which aggregating hyphae are embedded. Such matrix is found in rhizomorph, sclerotia and fruit bodies.

1.4 Ultra-structure of a fungal cell

Cell is bounded by a cell wall. Outermost membrane of cell is cell wall. Cell wall is made up of more than one layer and is made of fibrous structure chitin or cellulose or both. Inner to the cell wall is another layer known as plasma lemma. It is immediately attached to cytoplasm hence plasmalemma is also known as cytoplasmic membrance. Inside the plasmalemma is protoplasm. Protoplasm encloses nucleus, cytoplasm and cell inclusions.

Nucleus is most important organ in cell. It is eukaryotic and consists of nucleoplasm which is bounded by nuclear membrane. Nuclear membrane is not continuous, it is perforated. Inside the nucleoplasm nucleolus is suspended. Endoplasmic reticulum is seen in the cytoplasm. It is a small matrix which originates from nuclear membrane. It may be smooth or rough. If endoplasmic reticulum is studded with ribosome, it is rough endoplasmic, if not it is smooth endoplasmic reticulum. Ribosomes are proteinaceous bodies seen in the cytoplasm. Vacuoles are also present in cytoplasm, accumulation of metabolic products is seen in vacuoles. It is bounded by a membrane called tonoplast. Mitochondria help in respiration, size varies from 1 to 3μ. They are the sites of respiratory activities. Swollen portion of plasmalemma are called lomasomes. These are membraneous vesicular material embedded in plasmalemma. Inside the cell calcium oxalate crystals, glycogen, fat bodies, resins etc. are present.

1.4.1 Mycelium

An aggregation or network of hyphae constituting the filamentous thallus (body of the fungus) is known as mycelium. The protoplasm may be continuous in some fungi and in some other it may be interrupted at irregular intervals by partition which divides hyphae into cells. The partition or cross walls are called septa. (Septum [Latin] = partition). If walls are present such hyphae are known as septate hyphae. If walls are absent such hyphae are known as nonseptate or coenocytic hyphae or aseptate hyphae. The word coenocytic is derived from Greek word Koinose = comment, Kytos = a hollow vessel.

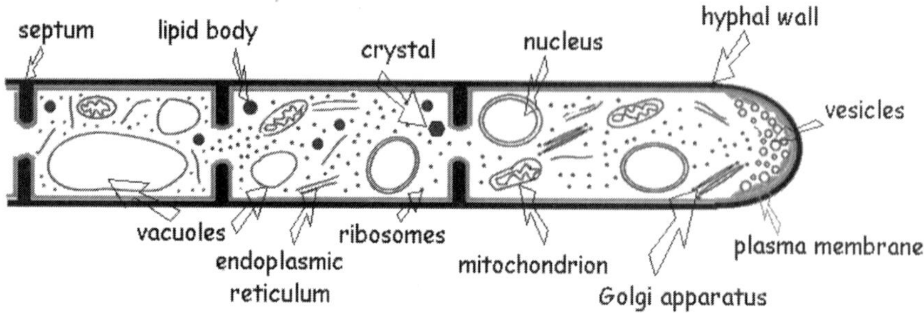

Fig 1.5. Fungal cell with organelles (Bowman and Free, 2006)

1.5 Septa

Septa are found in all filamentous fungi except Mastigomycotina and Zygomycotina. Several types of septa can be distinguished under the electron microscope. These are grouped as follows:

1.5.1. Solid septa: The simplest type of septum is a complete cross wall, produced by lower fungi to delimit old or reproductive parts of their mycelia.

1.5.2. Perforated septa: A variation of the above type is seen in the yeast like fungus.

Geotrichum has a simple septum perforated by a number of small bodies.

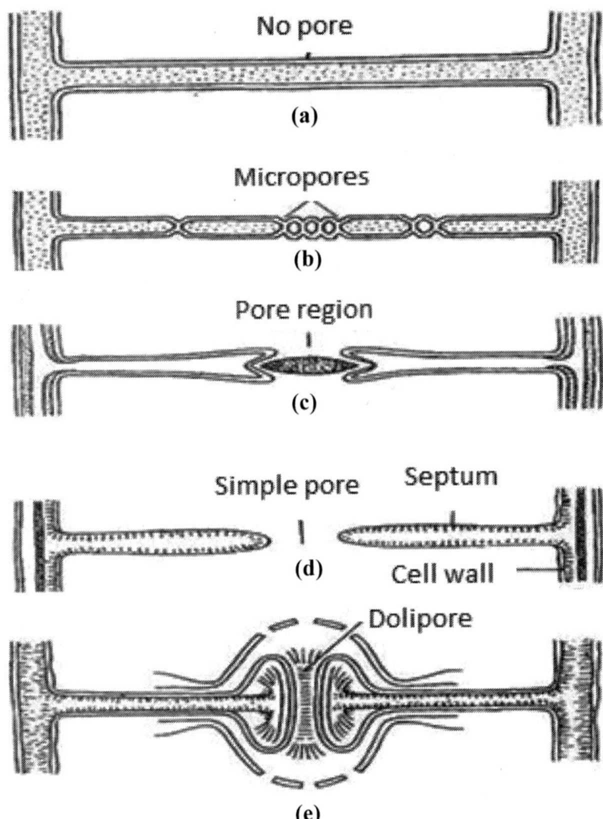

Fig 1.6. Different types of septa **(a)** Solid septum of lower fungi **(b)** Septa with micropore (*Geotrichum*) **(c)** Septa with simple pore (Deuteromycotina and Ascomycotina) **(d)** Septum of Trichomycetes **(e)** Dolipore septum (Basidiomycotina)

1.5.3 Ascomycetean type septa

This is the commonest type of septum found in most Ascomycotina and Deuteromycotina. This is a large central pore ranging from 0.05 μm in diameter. The septum grows inward from the wall as a diaphragm and leaves behind a small pore in the centre.

1.5.4 Bordered pit type septa

A very different appearance of the pore margin is characteristic of the septate Mucorales and Trichomycetes. The pore is surrounded by an overarching bifurcation of the septal margin. There is present a biumbonate, electron, dense plug of material blocking the pore but distinctly within plasmalemma.

1.5.5 Dolipore septa

This is a complex type of septum found in some stages in the life cycle of Holobasidiomycetes and Tremellales. The septum has a central pore but its margin has become inflated to form a swollen rim thereby making the pore more or less barrel shaped. In the septum the central electron lucent layer being sandwitched between two layers of electron opaque granular material. On either side of the pore, sheet of endoplasmic reticulum become closely associated to form a hemispherical cap, parenthosome which is perforated by pores.

1.6 Fungal Tissue

Large quantity of organized mycelium is known as plectenchyma (may be compactly or loosely woven). It is derived from Greek word Plekeih = to weave + enchyma = infusion i.e., a woven tissue.

1.6.1 Plectenchyma is of two types

1.6.1.1 Prosenchyma

The hyphae are loosely woven. Component hyphae do not lose their identity and tissue lie almost parallel to each other. These cells are seen in trama in *Agaricus*.

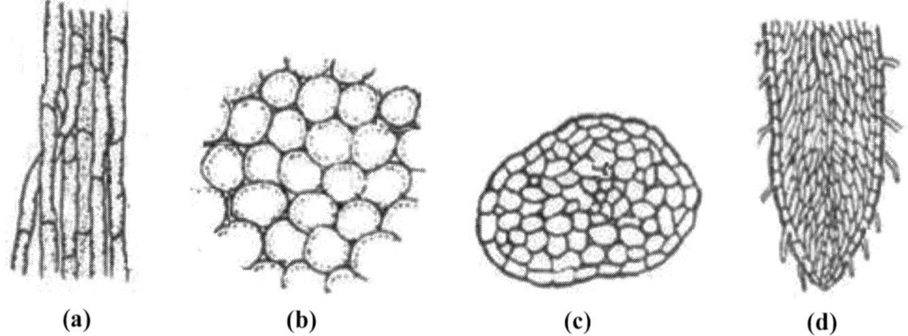

Fig.1.7. Different forms of fungal mycelium **(a)** Prosenchyma **(b)** Pseudoparenchyma **(c)** Sclerotium **(d)** Rhizomorphs.

1.6.1.2 Pseudoparenchyma

These are closely packed cells which are more or less isodiametric or oval in shape. It resembles the parenchyma of higher plants. Component hyphae lose their individuality and are not distinguishable. These cells are seen in sclerotium and rhizomorphs. Above tissue composes various types of vegetative and reproductive structures.

1.7 Modification of Mycelium

It absorbs nourishment (nutrients) and resists unfavorable conditions (extreme hot or extreme colds) for their survival. It serves as reproductive unit. Mycelial strands or cords consist of relatively undifferentiated aggregation of hyphae and are produced by a great variety of fungi. Mycelial strands may be employed by fungi which produce their fructification some distance away from the food base as in the stinkhorn, *Phallus impudicus*. Here the mycelial strand is more tightly aggregated and is referred as mycelial cord. The tip of the mycelial strands are capable of translocating nutrients and water in both direction. This property is important not only for decomposer fungi, but also for species forming mycorrhizal symbiosis with the roots of plants, many of which produce hyphal strands. Modifications of mycelium are as follows:

1.7.1 Rhizomorph

These are the rope like structures formed by more or less parallel or interwoven mass of highly differentiated tissues which adhere closely and are frequently cemented together. When the strands are loose they are called mycelial rind by their own inter weaning and anastomosing branches to form cord *ex. Sclerotium rolfsii*. But in *Armillaria*, a central core of larger, thin-walled elongated cells embedded in mucilage is surrounded by a rind of small, thick walled cells which are darkly pigmented due to melanin deposition in their walls. These root like aggregations are a means for *Armillaria* to spread underground from one tree root system to another. In nature, two kinds of rhizomorph are found- a dark cylindrical type and a paler, flatter type. The latter is particularly common beneath the bark of infected trees.

The most striking feature of the development of rhizomorphs is their compact growing point at the apex, which consists of small isodiametric cells protected by an apical cap of intertwined hyphae immersed in mucilage, which they produce. Behind the apex, there is a zone of elongation. The centre of the rhizomorph may be hollow or solid. Surrounding the central lumen or making up the central medulla is a zone of enlarged hyphae, 4-5 times wider than the vegetative hyphae. Toward the periphery of the rhizomorph, the cells become smaller, darker and thick walled. Extending outwards between the outer cells of the rhizomorph, there may be a growth of vegetative hyphae somewhat resembling the root hair zone in a higher plant. Rhizomorphs may develop on monokaryotic mycelia derived from single basidiospores, or on dikaryotic mycelia following fusion of compatible monokaryotic hyphae. Dikaryotic rhizomorphs of *Armillaria* do not posses clamp connections. Rhizomorphs are also produced by other Basidiomycotina and a few Ascomycotina.

1.7.2 Sclerotia

Sclerotia are pseudoparanchymatous aggregation of hyphae embedded in an extracellular glucan matrix. A hard melanized rind may be present or absent. Sclerotia serve as a survival structure and contain manitol, trehalose, glycogen, protein and lipid in certain species. They also serve as a reproductive role. They are produced by *Rhizoctonia, Sclerotinia, Sclerotium rolfsii* and *Claviceps* and *Botrytis cinerea.*

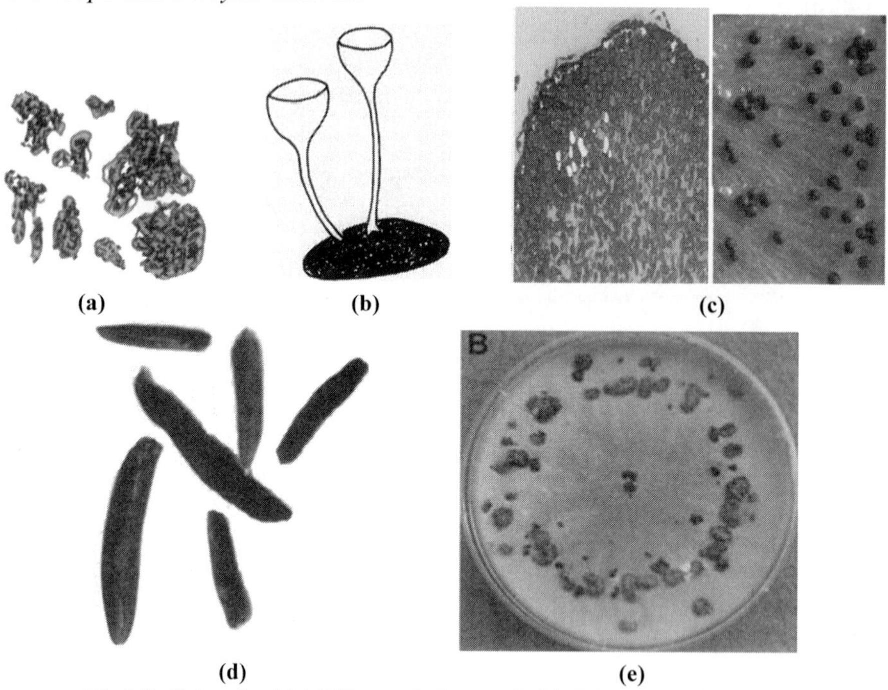

Fig.1.8. Sclerotia of **(a)** *Rhizoctonia bataticola* **(b)** *Sclerotinia sclerotiorum* **(c)** *Sclerotium rolfsii* **(d)** *Claviceps purpurea* **(e)** *Botrytis cinerea*

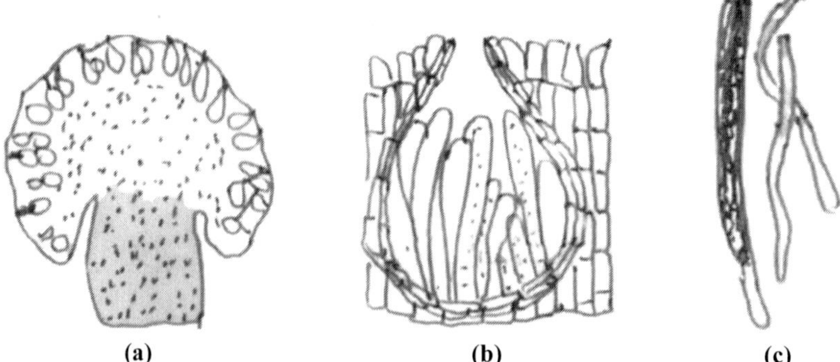

Fig. 1.9. (a) Perithecial stroma **(b)** Perithecium **(c)** Ascus and ascospores

Their form is variable from 100 µm to 200 mm or more in diameter. Majority are less than 5mm in diameter. The sclerotia of Australian *Polyporus mylittae* can reach the size of man's head (about 15 kg) and is known as native bread or black fellow's bread. Several kinds of development in sclerotia have been found..

I. Loose Type

These types of sclerotia are present in *Rhizoctonia solani.*, cause of black scurf of potato. Here the sclerotia size and shape is irregular but of uniform texture, brown or black, more or less loosely packed.The cells of hyphae are barrel shaped, anastomosing frequently. The branching is at right angles and pale brown to brown in colour. The mature sclerotium is made up of pseudoparenchymatous tissue. Towards the outside the hyphae are loosely arranged.

II. The Terminal Type

This type of sclerotia is produced by *Botrytis cinerea*, the cause of grey mold disease on a wide range of plants and by *Pyronema domesticum*. The apex of the hyphae leads by repeated dichotomous branching accomplished by cross wall formation. The hyphae then aggregate, melanized and produce mucilage, giving the appearance of solid tissue. The sclerotia consist of outer rind of thick walled cells and narrow cortex of thin walled pseudoparenchymatous cells with dense content and a medulla (large central part) of prosenparanchymatous cells. Nutrient reserves are stored in the cortical and medullary regions.

III. The Strand Type

Sclerotinia gladioli form sclerotia of this type. From main hyphae numerous thick side branches arise which lie parallel. These branches divided by septa into chain of short cells. These cells give rise to short branches, which lie parallel to parent hypha or grow out at right angles. The hyphae at the margin continue to branch. The mature sclerotium is differentiated into a rind of small thick walled cells. In *Sclerotium rolfsii* the mature sclerotium is differentiated into four zones, a thick cuticle, a rind made up of 2-4 layers of tangentially flattened cells, a cortex of thin walled cells with dense staining contents, and a medulla of loose filamentous hyphae with dense contents.

IV. Ergot type

The *Clavicep purpurea* form ergot like sclerotia. These sclerotia develop from a pre-existing mass of mycelium which fills and replaces the cereal ovary starting from the base and extending toward the apex. The outer layer forms a black ring enclosing colourless thick walled cells containing abundant storage lipids.

1.7.3 Pseudosclerotia

These are sclerotia like resting structure composed of fungal hyphae and some part of soil, stones etc., *ex. Polyporus mylittae*. The football like sclerotium is made up of white strata and translucent tissue. It has an outer smooth thin black rind. The tissue is made up of thin and thick walled hyphae which are abundant in white strata but sparse in translucent tissue and layered hyphae which occur only in translucent area. The translucent tissue may function as an extracellular nutrient and water. The detached sclerotia on germination form basidiocarp without wetting.

1.7.4 Pseudorrhiza

These are root shaped organs designed to unite the vegetative mycelium with the stipe of aerial basidiocarp *ex. Oudemansiella radicata* which grows attached to the tree stumps and buried wood and has a long tapering underground base to the stem.

1.7.5 Stroma

Stroma (stroma = bed, cushion): It is an aggregation of pseudoparenchymatous cells, in or on which spores are produced. It can be grouped into three categories.

1.7.5.1 Substratal stroma: It is of a diffuse or indefinite form. Its medulla is composed of loose hyphal network permeating and preserving as a food supply. In *Cercospora personata* below the epidermis cushion like structure is formed in the sub stomal region from which sporophores are produced.

1.7.5.2 Sclerotial stroma: Has a characteristic form and a strictly hyphal structure under the natural condition of its development. The stroma throughout its periphery has cavities in which spores are produced. These rounded structures are known as perithecial stroma *ex Claviceps purpurea*. Overwinter stroma produces stalked apothecia in spring.

1.7.5.3 Conidial stroma: In *Xylaria hypoxylon* stroma the powdery white conidia develop at the tips of the branches of conidial stroma and later asci develop in flask shaped perithecia at the base of the old stroma. The stroma is a food storage organ and is usually differentiated into two parts, a cortex of thick walled cells and a medulla of hyaline cells.

1.7.6 Haustoria

Houstorium is a special hyphal branch especially within the living cell of the host, for absorbtion of food. In *Albugo*, small spherical haustoria are present on intercellular mycelium. The haustorium is spherical or flattened, connected to the intercellular mycelium by a narrow stalk. Inside the haustoria plasma membrane, vesicles and tubules are formed by invagination of plasma membrane. The cytoplasm of haustorial tip is packed with mitochondria, ribosomes, endoplasmic reticulum and occasional lipid droplets. Nuclei may (*ex. Peronospora*) or may not be present in the haustoria of other Oomycetes. The base of the haustorium of *A.candida* is surrounded by a collar like sheath which is an extension of the host cell wall.

In *Phytophthora infestans* the haustoria appear as finger like protuberance. The haustorium is surrounded by outer haustorial matrix which is probably of fungal origin. This is delimited on the outside by host plasma membrane and on the inside by the wall and then the plasma membrane of the pathogen. Haustoria of *Phytophthora* do not normally contain nuclei. In *Blumeria graminis*, pathogen of grasses, wheat and cereals, the functional haustorium is formed by the enlarging tip of the penetration peg. The papilla remains as a collar around the peg, at the point of penetration of epidermal wall. The host plasma membrane is invaginated around the haustorium but it is not in direct contact with the plasma membrane of the haustorium. The haustorial wall is surrounded by the extra haustorial matrix, which is of host origin. It is a compartment with gelatinous matrix sealed by the host and pathogen plasma membrane and at the epidermal wall by a collar.

The dikaryotic haustorium of rust fungi is functionally very similar to that of Erysiphales. The haustorial mother cell emits a narrow penetration tube, which swells inside the host cell to form the haustorial body. The haustorial cytoplasm is surrounded by a haustorial membrane, the haustorial wall, the extra haustorial matrix and the extra haustorial membrane (*i.e.*, the modified

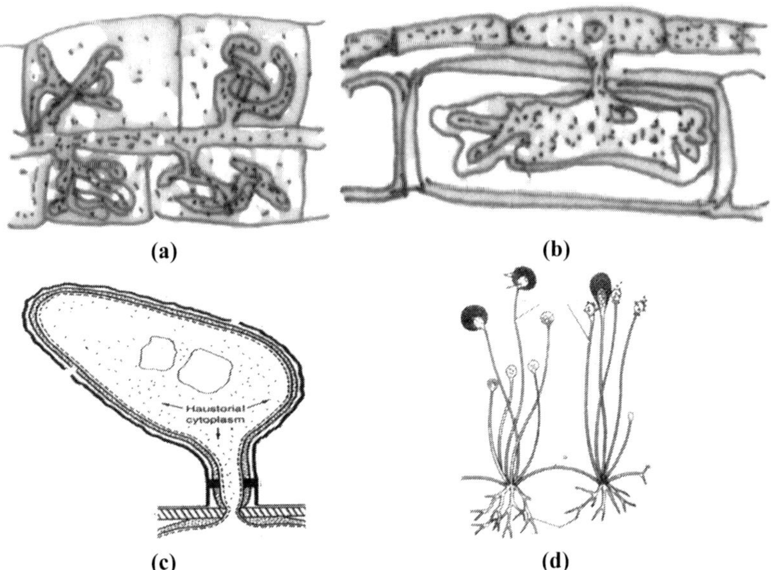

Fig. 1.10. Haustoria **(a)** *Erysiphe polygoni* **(b)** *Peronospora ficariae* **(c)** *Albugo candida* **(d)** *Rhizopus stolonifer*

plant plasmalemma). The haustorial matrix is sealed against the apoplast outside the infected plant cell by means of a neckband.

1.7.7 Rhizoids

It is a filamentous branched extension of a thallus acting as a feeding organ. In eucarpic monocentric thalli the rhizoids usually radiate from a single position on the sporangium wall but in polycentric forms a more extensive, branched rhizoidal system, the rhizomycelium, develops. In *Allomyces* the zoospore cyst produces at one point, a narrow germ tube which branches to form the rhizoidal system. In *Rhizopus* the rhizoids are present at the base of the sporangiophores which may grow in clusters and the stoloniferous habit. An aerial hypha grows out, and where it touches the substratum it bears rhizoids and sporangiophores.

1.7.8 Appressoria

It is a swelling on a germ tube or hypha, especially for attachment in an early stage of infection. In all powdery mildews except *Blumeria graminis*, a primary germ tube emerges and under suitable condition develops into an appresorium but in *B.graminis* the primary germ tube grows only to a limited distance. An appresorium is formed by unseparate secondary germ tube, which is much longer than the primary germ tube. Under optimum condition the germ tube elongates and its tip swell to form an appresorium, which is lobed and non- melanized.

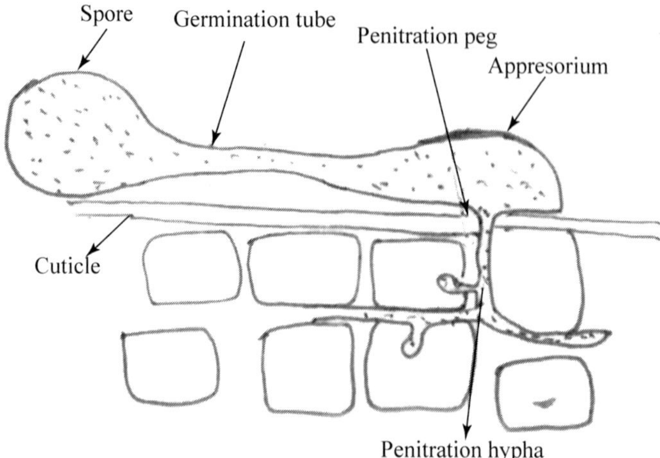

Fig. 1.11. Appressorium

2

Classification of Fungi

According to G.C. Ainsworth (1973) the fungus is free-living, parasitic or mutualistic, saprotrophs, necrotrophs or biotrophs and devoid of chlorophyll. Cell wall composition very variable, majority contain chitin and glucans, reserve materials glycogen, oil and mannitol, characteristic disaccharide and trehalose. Some yeast-like but majority with thread-like filaments, hyphae, branched profusely to form the vegetative mycelium on which spores are produced, asexually or sexually, free on hyphae or enclosed in complex reproductive structures. Separated mainly on morphology of the latter.

2.1 Divisions: Myxomycota

These are wall-less and quite unusual organisms only included in the fungi as mostly studied by mycologists. Possess either a plasmodium, a mass of naked, multinucleate protoplasm, which feeds any ingesting particulate matter and moves by amoeboid cells. Both of slimy consistencies hence slime molds. Four classes:

2.1.1 Class: Acrasiomycetes (Cellular slime molds)

Assimilative phase, free living amoebae which aggregate to form a pseudoplasmodium before reproduction.

Order	Family	Genus
2.1.1.1 Protosteliales (2F)	1. Protosteliaceae	*Protostelium*
	2. Cavosteliaceae	*Cavostelium*
2.1.1.2 Acrasiales (3F)	1. Acrasiaceae	*Acrasis*
	2. Guttulinaceae	*Guttulinopsis*
	3. Copromyxaceae	*Copromyxa*
2.1.1.3 Dictyosteliales (2F)	1. Dyctyosteliaceae	*Dictyostelium*
	2. Acytosteliaceae	*Acytostelium*

2.1.2 Class Hydromyxomycetes (Net slime molds)

Spindle-shaped cells with slimy filaments that join together to form a slimy network; mostly parasitic on marine plants.

Order

2.1.2.1 Hydromyxales

2.1.2.2 Labyrinthulales (2F)

1. Labyrinthulaceae	*Labyrinthula*
2. Thraustochytriaceae	

2.1.3 Class Myxomycetes (True slime mold)

Assimilative phase, a true plasmodium (*i.e.*, multinucleate protoplasmic mass), free-living, saprophytic.

2.1.3.1 Ceratiomyxales (1F)	1. Ceratiomyxaceae	*Ceratiomyxa*
2.1.3.2 Physarales (2F)	1. Physaraceae	*Physarum*
	2. Didymiaceae	
2.1.3.3 Liceales (3F)	1. Liceaceae	*Licea, Lycogala*
	2. Reticulariaceae	
	3. Cribrariaceae	
2.1.3.4 Echinosteliales (2F)	1. Echinosteliaceae	*Echinostelium*
	2. Clastodermataceae	
2.1.3.5 Trichiales (2F)	1. Dianemaceae	
	2. Trichiaceae	*Trichia*
2.1.3.6 Stemonitales (1F)	1. Stemonitaceae	*Stemonitis*

2.1.4 Class Plasmodiophoromycetes (Endoparastic slime molds)

It is included in both Myxomycota and Eumycota. Ainsworth (1973) included them as a class of the Mastigomycotina (Zoosporic fungi). Small plasmodia, parasitic within cells of fungi, algae or higher plants.

2.1.4.1 Plasmodiophorales (1F)	1. Plasmodiophoraceae	*Plasmodiophora*
		Spongospora.
		Woronina, Polymyxa.

Class Acrasiomycetes and Hydromyxomycetes are perhaps more appropriately considered as protozoa, although studied by mycologists.

2.2 Division Eumycota

True fungi, all with walls. Customary to recognize five sub divisions:

2.2.1 Sub division Mastigomycotina: Motile cells-zoospores present; perfect-state spore-oospores. Three classes:

2.2.1.1 Class Chytridiomycetes: Often unicellular, zoospores have single, posterior, whiplash flagellum.

2.2.1.1.1 Chytridiales (9F)	1. Olpidiaceae	*Olpidium, Rozella*
	2. Achlyogetonaceae	*Achlyogeton*
	3. Synchytriaceae	*Synchytrium*
	4. Phlyctidiaceae	*Rhizophydium*
	5. Rhizidiaceae	*Polyphagus, Rhizophlyctis*
	6. Cladochytriaceae	*Cladochytrium*
	7. Physodermataceae	*Physoderma*

	8. Chytridiaceae	*Chytridium* *Chytriomyces,*
	9. Megachytridiaceae	*Nowakowskiella*
2.2.1.1.2 Harpochytriales (1F)	1. Harpochytriaceae	*Harpochytrium*
2.2.1.1.3 Blastocladiales (3F)	1. Coelomomycetaceae	*Coelomomyces*
	2. Catenariaceae	*Catenaria*
	3. Blastocladiceae	*Allomyces,* *Blastocladiella*
2.2.1.1.4 Monoblepharidales (2F)	1. Monoblepharidaceae	*Monoblepharis*
	2. Gonapodyaceae	*Monoblepharella,* *Gonapodya*

2.2.1.2 Class Hyphochytridiomycetes: Often unicellular, zoospores have single, anterior, tinsel flagellum.

2.2.1.2.1 Hyphochytriales (3F)	1. Anisolpidiaceae	*Anisolpidium,* *Canteriomyces*
	2. Rhizidiomycetaceae	*Rhizidiomyces,* *Rhizidiomycopsis*
	3. Hyphochytriaceae	*Hyphochytrium.*

2.2.1.3 Class Oomycetes: Usually mycelia (aseptate), zoospores have two flagella, one directed back-wards, of whiplash type and one forward tinsel-type.

2.2.1.3.1 Saprolegniales (1F)	1. Saprolegniaceae	*Saprolegnia, Achlya,* *Aphanomyces* *Thraustotheca*
2.2.1.3.2 Lagenidiales (3F)	1. Lagenidiaceae	*Lagenidium*
	2. Olpidiopsidaceae	*Olpidiopsis,* *Rozellopsis,*
	3. Sirolpidiaceae	*Lagenisma,* *Pontisma,* *Sirolpidium*
2.2.1.3.3 Leptomitales (2F)	1. Leptomitaceae	*Leptomitus*
	2. Rhipidiaceae	
2.2.1.3.4 Peronosporales (3F)	1. Pythiaceae	*Pythium,* *Phytophthora,* *Sclerophthora*
	2. Peronosporaceae	*Peronospora,* *Plasmopara* *Sclerospora, Bremia*
	3. Albuginaceae	*Albugo*

2.2.2 Sub-division Zygomycotina

Usually mycelia aseptate, asexual spores non-motile, formed inside a sporangium; perfect-state spores-zygospores. Two classes:

2.2.2.1 Class Zygomycetes: Usually saprophytic, parasitic or predacious, mycelium immersed in host tissue.

Order	Family	Genus
2.2.2.1.1 Mucorales (14F)	1. Mucoraceae	*Mucor, Rhizopus, Absidia Phycomycea, Zygorhynchus.*
	2. Pilobolaceae	*Pilobolus, Pilaira*
	3. Mortierellaceae	*Mortierella, Aquamortierella.*
	4. Saksenaeaceae	*Saksenaea.*
	5. Endogonaceae	*Endogone*
	6. Radiomycetaceae	*Radiomyces.*
	7. Thamnidiaceae	*Thamnidium, Chaetocladium*
	8. Choanephoraceae	*Choanephora, Gibberella, Blakeslea*
	9. Cunninghamellaceae	*Cunninghamella*
	10. Helicocephalidaceae	*Helicocephalum*
	11. Syncephalastraceae	*Syncephalastrum*
	12. Piptocephalidaceae	*Piptocephalis, Syncephalis*
	13. Dimargartiaceae	*Dimargaris, Dispira*
	14. Kickxellaceae	*Kickxella, Coemansia*
2.2.2.1.2 Entomophthorales (1F)	1. Entomophthroceae	*Entomophthora, Basidiobolus*
2.2.2.1.3 Zoopagales (2F)	1. Zoopagaceae	*Zoopage, Stylopage.*
	2. Cochlonemaceae	*Cochlonema*

2.2.2.2 Class Trichomycetes: Often parasitic on arthropods, mycelium not immersed in host tissue.

Order	Family	Genus
2.2.2.2.1 Harpellales (2F)	1. Harpellaceae	*Harpella*
	2. Genistellaceae	*Genistellospora*
2.2.2.2.2 Asellariales (1F)	1. Asellariaceae	*Asellaria*
2.2.2.2.3 Eccrinales (3F)	1. Palavasciaceae	*Palavascia*
	2. Parataeniellaceae	*Parataeniella*
	3. Eccrinaceae	*Enterobryus*
2.2.2.2.4 Amoebidiales (1F)	1. Amoebidiaceae	*Amoebidium*

2.2.3 Sub-division Ascomycotina

Yeast or septate mycelium; asexual spores non-motile, not formed inside sporangium; perfect- state spores-ascospores, formed in an ascus usually within a fruit body or ascocarp. Six classes:

2.2.3.1 Class Hemiascomycetes: No ascocarps and ascogenous hyphae present; thallus yeast-like or mycelia.

Order	Family	Genus
2.2.3.1.1 Protomycetales (1F)	1. Protomycetaceae	*Protomyces, Protomycopsis*
2.2.3.1.2 Endomycetales (4F)	1. Ascoideaceae	*Ascoidea, Dipodascus*
	2. Spermophthoraceae	*Spermophthora, Ashbya*
	3. Endomycetaceae	*Endomyces, Eremascus*
	4. Saccharomycetaceae	*Saccharomyces*
2.2.3.1.3 Taphrinales (1F)	1. Taphrinaceae	*Taphrina*

2.2.3.2 Class Loculoascomycetes: Ascocarp and ascogenous hyphae present; thallus mycelia; asci bitunicate (2-walled), fruit body an ascostroma i.e., a mass of tissue with locules.

Order	Family	Genus
2.2.3.2.1 Myriangiales (4F)	1. Myriangiaceae	*Elsinoe, Myriangium*
	2. Saccardiaceae	*Piedraia*
	3. Saccardinulaceae	
	4. Atichiaceae	
2.2.3.2.2 Dothidiales (8F)	1. Trichothiriaceae	
	2. Chaetothyriaceae	
	3. Pseudosphaeriaceae	
	4. Parodiopsidaceae	
	5. Englerulaceae	
	6. Dothioraceae	
	7. Capnodiaceae	*Capnodium, Mycosphaerella*
	8. Dothideaceae (Mycosphaerellaceae)	*Guignardia*
2.2.3.2.3 Pleosporales (8F)	1. Dimeriaceae	
	2. Venturiaceae	*Venturia*
	3. Mesnieraceae	
	4. Botryosphaeriaceae	

	5. Lophiostomataceae	*Lophiostoma*
	6. Sporomiaceae	
	7. Pleosporaceae	*Pleospora, Cochliobolus*
	8. Mycoporaceae	
2.2.3.2.4 Hysteriales (6F)	1. Hysteriaceae	
	2. Arthoniaceae	
	3. Opegraphaceae	
	4. Phillipsiellaceae	
	5. Patellariaceae	
	6. Lecanactidaceae	
2.2.3.2.5 Hemisphaeriales (11F)	1. Micropeltidiaceae	
	2. Munkiellaceae	
	3. Microthyriaceae	*Microthyrium*
	4. Trichopeltinaceae	
	5. Parmulariaceae	
	6. Aulographaceae	
	7. Asterinaceae	
	8. Brefeldiellaceae	
	9. Leptopeltidaceae	
	10. Stephanotheceae	
	11. Schizothyriaceae	

2.2.3.3 Class Plectomycetes: Ascocarp and ascogenous hyphae present; thallus mycelia; asci unitunicate

Order	Family	Genus
2.2.3.3.1 Gymnoascales (3F)	1. Dendrosphaeriaceae	*Dendrosphaera*
	2. Gymnoascaceae	*Ajellomyces, Nannizzia*
	3. Onygenaceae	*Onygena*
2.2.3.3.2 Eurotiales (3F)	1. Cephalothecaceae	*Cephalotheca*
	2. Pseudoeurotiaceae	*Pseudoeurotium, Pleuroascus*
	3. Eurotiaceae	*Byssochlamys, Eurotium*
2.2.3.3.3 Ascosphaerales (1F)	1. Ascophaeraceae	*Ascosphaera*
2.2.3.3.4 Elaphomycetales (1F)	1. Elaphomycetaceae	*Elaphomyces*
2.2.3.3.5 Microascales (1F)	1. Microascaceae	*Microascus*
2.2.3.3.6 Ophiostomatales (1F)	1. Ophiostomataceae	*Ceratocystis*

(one walled), evanescent *i.e.*, breakdown at maturity, scattered within a closed fruit body-cleistothecium.

2.2.3.4 Class Laboulbeniomycetes: Ascocarp and ascogenous hyphae present; thallus reduced; asci regularly arranged within the ascorarp (a perithecium); asci unitunicate; inoperculate; exoparasites of arthropods.

2.2.3.4.1 Laboulbeniales (4F)	1. Ceratomycetaceae	
	2. Euceratomycetaceae	
	3. Herpomycetaceae	
	4. Laboulbeniaceae	*Laboulbenia*
		Stigatomyces
2.2.3.4.2 Spathulosporales (1F)	1. Spathulosporaceae	*Spathulospora.*

2.2.3.5 Class Pyrenomycetes: Ascocarp and ascogenous hyphae present; thallus reduced, asci unitunicate; regularly arranged within the ascocarp (a perithecium); asci inoperculate with apical pore or slit.

Order	Family	Genus
2.2.3.5.1 Erysiphales (1F)	1. Erysiphaceaae	*Erysiphe, Phyllactinia*
		Uncinula,
		Sphaerotheca
		Microsphaera,
		Podosphaera
		Oidium, Ovulariopsis
2.2.3.5.2 Meliolales (1F)	1. Meliolaceae	*Meliola*
2.2.3.5.3 Sphaeriales (7F)	1. Chaetomiaceae	*Chaetomium*
	2. Xylariaceae	*Xylaria, Rosellinia*
	3. Diatripaceae	
	4. Phyllachoraceae	*Phyllachora*
	5. Coniochaetaceae	
	6. Sordariaceae	*Sordaria,*
		Neurospora
		Podospora.
	7. Polystigataceae	*Glomerella*
2.2.3.5.4 Clavicipitales (1F)	1. Clavicipitaceae	*Claviceps,*
		Cordyceps.
2.2.3.5.5 Diaporthales (3F)	1. Melanosporace	*Melanospora*
	2. Gnomoniaceae	*Gnomonia*
	3. Diaporthaceae	*Diaporthe, Endothia*
2.2.3.5.6 Hypocreales (2F)	1. Hypocreaceae	*Nectria, Hypocrea,*
		Gibberella
	2. Hypomycetaceae	*Hypomyces*
2.2.3.5.7 Coronophorales (1F)	1. Coronophoraceae	*Coronophora*
2.2.3.5.8 Coryneliales (1F)	1. Coryneliaceae	*Coryneli*

2.2.3.6 Class Discomycetes: Ascocarp and a genus asco hyphae present; thallus mycelia; asci unitunicate, regularly arranged in a cup-shaped ascocarp (an apothecium); asci operculate with apical pore.

2.2.3.6.1 Medeolariales (1F)	1. Medeolariaceae	*Medeolaria*
2.2.3.6.2 Cyttariales (1F)	1. Cyttariaceae	*Cyttaria*
2.2.3.6.3 Tuberales (3F)	1. Tuberaceae	Tuber
	2. Terfeziaceae	*Terfezia*
	3. Geneaceae	*Genea*
2.2.3.6.4 Pezizales (5-13F)	1. Thelebolaceae	*Sarcosoma, Cockeina, Thindia*
	2. Sareoseyphaceae	
	3. Pezizaceae	*Peziza*
	4. Ascobolaceae	*Ascobolus*
	5. Pyronemataceae	*Pyronema*
	6. Morchellaceae	*Morchella, Verpa*
	7. Helvellaceae	*Helvella*
2.2.3.6.5 Rhytismatales (2F)	1. Rhytismataceae	*Rhytisma*
	2. Hypodermataceae	*Lophodermium*
2.2.3.6.6 Ostropales (1-F)	1. Stictidaceae	
2.2.3.6.7 Helotiales (10-F)	1. Ascocorticiaceae	
	2. Orbiliaceae	
	3. Dermatiaceae	*Diplocarpon*
	4. Hyaloscyphaceae	
	5. Sclerotiniaceae	*Sclerotinia*
	6. Helotiaceae	
	7. Geoglossaceae	*Geoglossum*
	8. Hemiphacidiaceae	
	9. Gelatinodiscaceae	
	10. Phacidiaceae	

2.2.4 Sub. Division Basidiomycotina

Yeast or septate mycelium; asexual spores absent; if present non-motile; perfect-state spores-basidiospores formed on a basidium.

2.2.4.1 Class Teliomycetes: Basidiocarp lacking teliospores (encysted probasidia) grouped in sori or scattered within the host tissue; parasitic on vascular plants.

2.2.4.1.1 Uredinales (3F)	1. Pucciniaceae	*Puccinia, Gymnosporangium, Uromyces, Phragmidium, Ravenelia, Hemileia*
	2. Melampsoraceae	*Melampsora, Cronartium, Urediniopsis, Masseeella,*
	3. Coleosporaceae	*Coleosporium.*

2.2.4.1.2 Ustilaginales (3F)	1. Graphiolaceae	*Graphiola.*
	2. Ustilaginaceae	*Ustilago, Sphacelotheca Tolyposporium, Sorosporium*
	3. Tilletiaeceae	*Tilletia, Neovossia, Urocystis, Entyloma*

2.2.4.2 Class Hymenomycetes: Basidiocarp present, basidia typically arranged in organized layer, hymenium which is completely or partly exposed at maturity, basidia septate or aseptate; basidiospores ballistospores.

2.2.4.2.1 Sub Class Phragobasidiomycetidae

2.2.4.2.1.1 Tremellales (3F)	1. Sirobasidiaceae	*Sirobasidium*
	2. Hyaloriaceae	
	3. Tremellaceae	*Tremella, Exidia*
2.2.4.2.1.2 Auriculariales (1F)	1. Auriculariaceae	*Helicobasidium, Auricularia*
2.2.4.2.1.3 Septobasidiales (1F)	1. Septobasidiaceae	*Septobasidium, Uredinella.*

2.2.4.2.2 Sub Class Holobasidiomycetidae

2.2.4.2.2.1 Exobasidiales (1F)	1. Exobasidiaceae	*Exobasidium*
2.2.4.2.2.2 Dacrymycetales (1F)	1. Dacrymycetaceae	*Dacrymyces*
2.2.4.2.2.3 Brachybasidiales (1F)	1. Brachybasidiaceae	*Brachybasidium*
2.2.4.2.2.4 Tulasnellales (2F)	1. Tulasnellaceae	*Tulasnella*
	2. Ceratobasidiaceae	*Ceratobasidium, Thanetophorus*
2.2.4.2.2.5 Aphyllophorales (22F)	1. Cantharellaceae	12. Fistulinaceae
	2. Coniophoraceae	13. Ganodermataceae
	3. Gomphaceae	14. Polyporaceae
	4. Stereaceae	15. Corticiaceae
	5. Schizophyllaceae	16. Punctulariaceae
	6. Clavariaceae	17. Thlephoraceae
	7. Sparassidaceae	18. Cyphelliaceae
	8. Auriscalpiaceae	19. Clavunilaceae
	9. Echinodorntiaceae	20. Bankariaceae
	10. Hydnaceae	21. Hericiaceae
	11. Bondazzewiaceae	22.Hymenochaetaceae
2.2.4.2.2.6 Agaricales (15F)	1. Boletaceae	*Boletus*
	2. Hygrophoraceae	*Hygrocybe.*
	3. Tricholomataceae	*Armillaria, Pleurotus*
	4. Entolomataceae	*Rhodocybe*
	5. Pluteaceae	*Volvariella*
	6. Lepiotaceae	*Lepiota*
	7. Agaricalceae	*Agaricus, Amanita.*

	8. Bolbitiaceae	
	9. Strophariaceae	*Psilocybe*
	10. Copriniaceae	*Coprinus*
	11. Cortinariaceae	*Cortinarius*

2.2.4.3 Class Gasteromycetes: Basidiocarp present; basidia arranged in hymenium, enclosed within the basidiocarp;basidia aseptate.

2.2.4.3.1 Podaxales (3F)	1. Podaxaceae	*Podaxis*
	2. Cribbeaceae	
	3. Secotiaceae	
2.2.4.3.2 Hymenogastrales (8F)		
2.2.4.3.3 Sclerodermatales (4F)	1. Sclerodermataceae	*Scleroderma*
	2. Glischrodermataceae	
	3. Broomeriaceae	
	4. Astraeaceae	
2.2.4.3.4 Lycoperdales (4F)	1. Arachiniaceae	
	2. Mesophilliaceae	
	3. Geastraceae	*Geastrum, Trichaster*
	4. Lycoperdaceae	*Lycoperdon, Calvatia*
2.2.4.3.5 Phallales (6F)	1. Hysterangiaceae	
	2. Clathraceae	
	3. Phallaceae	*Phallus, Clathrus,*
	4. Claustulaceae	
	5. Protoplallaceae	
	6. Gelopellidareceea	
2.2.4.3.6 Nidulariales (2F)	1. Sphaerobolaceae	*Sphaerobolus*
	2. Nidulariaceae	*Nidularia, Cyathus*

2.2.5 Sub-division Deuteromycotina

Yeast or saptate mycelium; asexual spores as in Ascomycotina; perfect-state spores absent, rare or unknown. Three classes:

2.2.5.1 Class Hyphomycetes: Mycelial; sterile or bearing asexual spores directly on hyphae or on special branches, conidiophores.

2.2.5.1.1 Hyphomycetales (2F)	1. Moniliaceae	*Aspergillus* *Penicillium* *Gliocladium* *Trichoderma* *Verticillium* *Botrytis, Sporothrix* *Geotrichum* *Cercospora* *Trichophyton* *Microsporum*

	2. Dematiaceae	*Alternaria* *Curvularia* *Drechslera*
2.2.5.1.2 Stilbellales (1F)	1. Stilbellaceae	*Trichurus, Graphium* *Isaria, Coremium*
2.2.5.1.3 Tuberculariales (1F)	1. Tuberculariaceae	*Fusarium* *Tubercularia* *Volutella* *Epicoccum* *Exosporium.*
2.2.5.1.4 Agonomycetales (1F)	1. Agonomycetaceae	*Rhizoctonia* *Sclerotium*

2.2.5.2 Class Blastomycetes: Budding (yeast or yeast-like) cells with or without pseudomycelium; true mycelium lacking or not well-developed.

2.2.5.2.1 Sporobolomycetales (1F)	1. Sporobolomycetaceae	*Bullera,* *Sporobolomyces*
2.2.5.2.2 Cryptococales (1F)	1. Cryptococcaceae	*Cryptococcus* *Candida* *Rhodotorula* *Trichospora*
	3. Leptostromataceae	*Leptostroma* *Discosia.*
	4. Excipulaceae	*Excipula, Discula.*

2.2.5.3 Class Coelomycetes: Mycelial; asexual spores formed in a flask-shaped structure pycnidium or on a 'pad' of fungal tissue-acervulus.

2.2.5.3.1 Order Sphaeropsidiales (4F)	1. Sphareopsidiaceae	*Phyllosticta, Phoma,* *Phomopsis* *Macrophomina* *Coniothyrium.* *Chaetomella* *Ascochyta, Darluca.* *Diplodia* *Botrydiplodia* *Septoria*
	2. Nectrioidaceae	*Zythia*
2.2.5.3.2 Melanconiales (1F)	1. Melanconiaceae	*Colletotrichum* *Gloeosporium* *Pestalotia* *Pestalotiopsis* *Monochaetia* *Melanconium*

3

Fungal Nutrition

Fungi are achlorophyllus so they have to depend on other organisms hence heterotrophic nutrition. Depending on the nutrition, fungi are classified as Saprophytes, Parasites and Symbionts.

3.1 Saprophyte (sapros = rotten; phyton = plant)

Fungi which get their nutrition only from dead organic are called saprophytes *ex*: *Saprolegnia, Rhizopus, Mucor, Alternaria* etc.

3.1.1 Facultative saprophyte (Facultas = ability)

Ability of a fungus to become saprophyte. It is a parasite *ex. Ustilago*

3.1.2 Obligate saprophyte (obligate = to bind itself)

When they strictly get their food only from dead organic matter and can never become parasites are called obligate saprophyte *ex. Mucor.*

3.2 Parasites

These get their nutrition from other living organisms. Lead as parasite throughout the life or part of life. An organism on which they parasitize is called as host. When they damage the host, they are called as pathogens. All the pathogens need not be a parasite.

3.2.1 Facultative Parasites

Ability of a fungus to become parasitic.Actually, it is a saprophyte. *ex. Pythium aphanidermatum.*

3.2.2 Obligate parasite

Parasite which get its nutrition only from living tissues *ex*. Rust fungi (*Puccinia graminis var tritici, Albugo, Peronospora, Synchytrium* etc.

3.3 Symbionts

Many fungi are involved in close association, known as symbiotic associations, which are mutually beneficial to both organisms.

Lichens and mycorrhizas associations have enabled some photosynthetic organisms to colonise deserted environments (*Usnea*, *Glomus*).

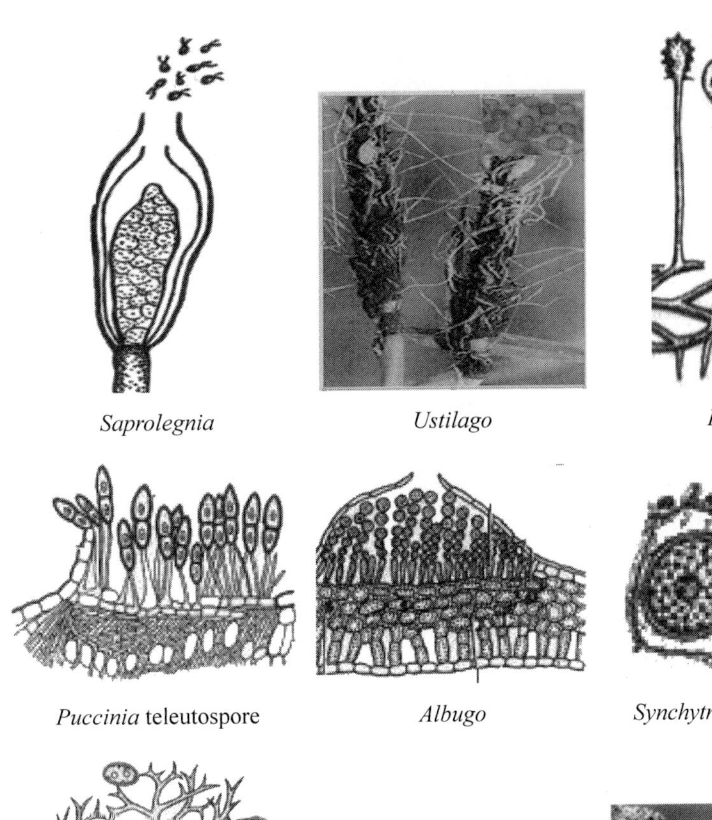

Saprolegnia *Ustilago* *Rhizopys*

Puccinia teleutospore *Albugo* *Synchytrium* sporangium

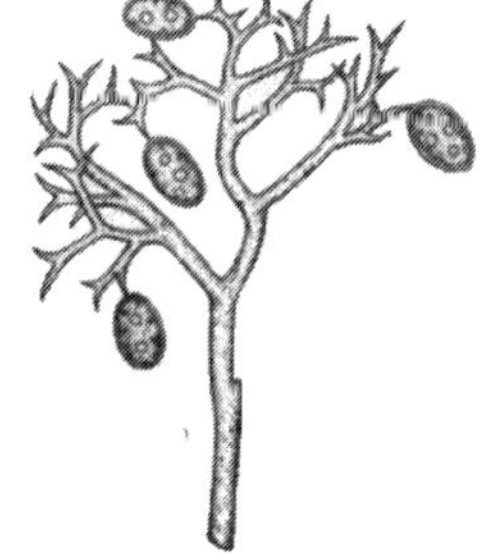

Peronospora sporangium

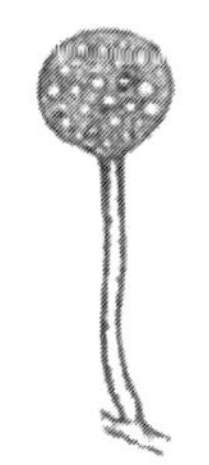

Pythium

Lichen

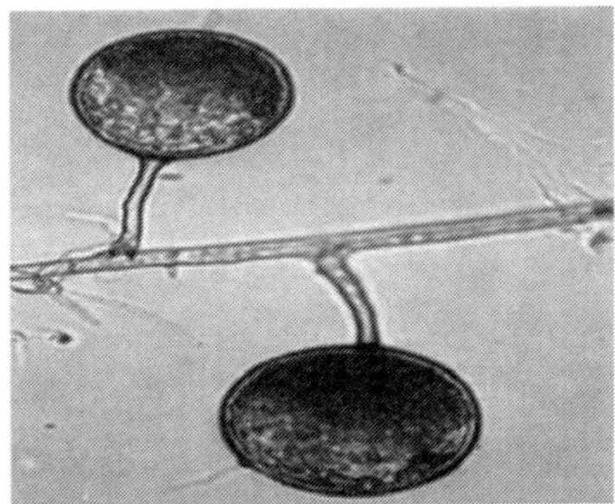

Glomus

4

Reproduction in Fungi

Reproduction in fungi is the formation of new individuals having all the characters of species. It is of two types. Asexual, somatic, or vegetative reproduction does not involve the union of nuclei of sex cells or sex organs. Sexual reproduction is characterized by the union of two opposite nuclei. During reproduction entire thallus is converted into one or more reproductive units. It is called holocarpic fungi (holo = whole, korpos = fungi) *ex. Plasmodiophora* and *Synchytrium*. In *Plasmodiophora* vegetative body is plasmodium, which is transformed into reproductive structure. In some fungi only a part of thallus is transformed into one or more reproductive structures. (eu = true, good, carpus = fruit). These fungi are known as eucarpic. Both somatic and reproductive structures coexist. *Ex. Rhizopus, Puccinia* and *Helminthosporium.*

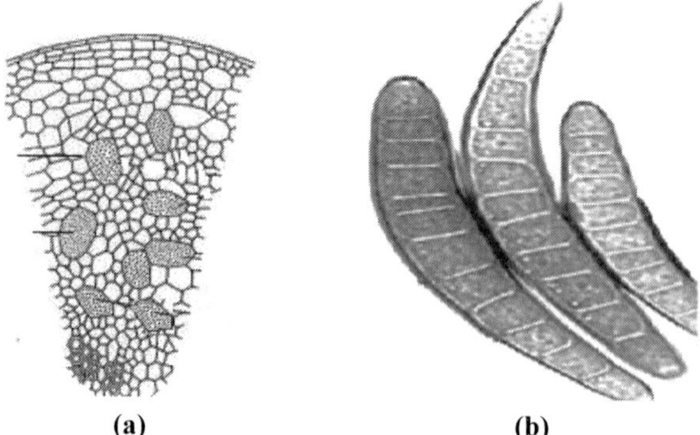

Fig. 4.1. **(a)** Holocarpic fungi (*Plasmodiophora sp*) **(b)** Eucarpic fungi (*Helminthosporium* sp)

4.1 Importance

The two methods of reproduction in fungi *i.e.*, asexual and sexual are intended to accomplish more or less two distinct acts in the life history of the fungus. The asexual spores are primarily meant for rapid dissemination while the sexual fruit mainly helps the fungus to survive through adverse conditions.

Asexual spores are essentially the main dispersive units and is also known as imperfect stage or anamorphic stage. They are produced in enormous quantities to compensate for the loss during dispersal due to natures adversities. They are provided with quick mechanism for quick release and transport at the appropriate time and are capable of immediate germination to establish the individual at the new sites. Such units have the ability to retain their power of germination during the process of long-distance dispersal.

Fungi multiply by means of their propagules. The unit is capable to grow into new thallus. In most fungi, these propagules are differentiated spores which are the basic unit of reproduction. The simple or branched spore bearing hyphae are known as sporophores but in some fungi spores may be formed directly by the hyphal cell (*eg.* chlamydospores) without formation of a sporophore. Simple or filamentous sporophores are known as sporangiophore (*ex.* Oomycetes and Zygomycetes) and conidiophores (Deuteromycetes). Compound sporophores are aggregation of hypha from somatic structures.

4.2 Types of spore fruit

The sporophores bear fruit bodies or form fructifications which may be asexual *ex.* Plasmodiophoromycetes, Chitridiomycetes, Oomycetes and Zygomycetes.

The asexual spores are usually enclosed in a simple sac. These are called sporangiospores or zoosporangiospores, depending upon the spores contained in them. In higher form of fungi, however, there is a tendency to organize complex aggregates of spore bearing hypha frequently surrounded by more or less protective tissues.

4.2.1 Synnema

A group of conidiophores often unites at the base and part way up toward the tip and form a structure known as synnema *ex. Penicillium claviforme* and *Cephalotrichum stemonitis*. The various types of synnema are present viz simple, compound, some made up of parallel conidiophores and others, where the hyphae making up the synnema are intricately interwoven. The top of synnema is often much branched. The conidia arise at the tips of the numerous branches, *ex. Arthrobotryum, Penicilium claviforme.*

4.2.2 Stroma

It is a compact somatic structure made up of pseudoparenchymatous cells on or in which fructifications are usually formed. In *Daldinia* sp. the stromata is composed entirely of fungal tissue in which perithecia with definite walls are embedded, with their necks slightly protrude. In *Xylaria hypoxylon* the powdery white conidia develop at the tips of the branches of the conidial stroma and later asci develop in flask shaped perithecia at the base of the old stroma.

4.2.3 Sporodochia

It is a fruiting body formed by cushion like aggregation of hyphae, which breaks through the host surface and bear conidiophores. These structures may also be formed in the mass of hyphae lying superficially over a substrate *ex. Fusarium lini.*

4.2.4 Acervulus

It is a saucer shaped fructification which may develop inside host tissue or may be superficial. Subepidermal acervuli develop from a pseudoparenchymatous stroma and as the acervulus matures, the host epidermis becomes ruptured to expose conidia formed from conidiogenous cells lining the base of the saucer. The conidia are held together in slime and are chiefly disperesed by rain splash *ex. Colletotrichum.*

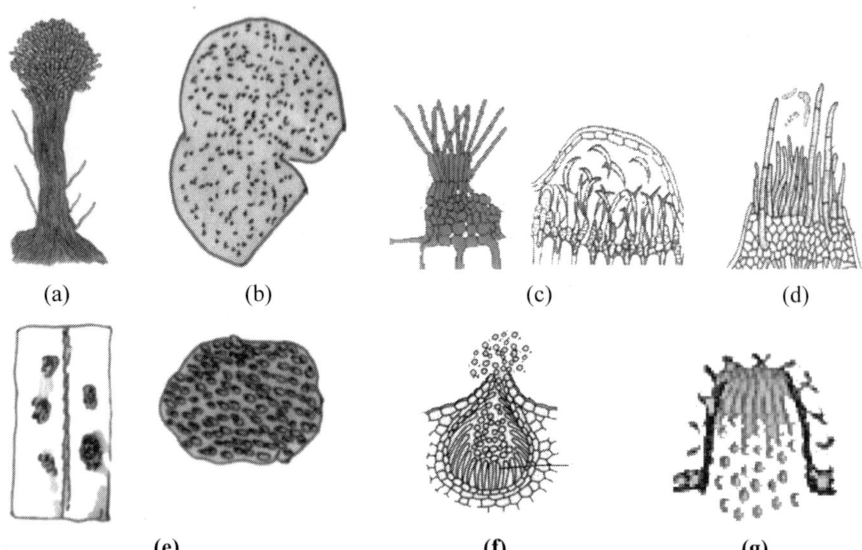

Fig. 4.2. Types of fruiting bodies **(a)** Synnema **(b)** Stroma **(c)** Sporodochia **(d)** Acervulus **(e)** Sorus **(f)** Pycnidium **(g)** Aecium

4.2.5 Sorus

The sporophores and spores are grouped into small or large masses or clusters *ex.* white blister in white rust and smut. A single sorus is an aggregation of spores along or mixed with spore bearing hyphae and sometimes sterile filament or cells. It may be covered with layers of host cell or may be naked. The spores and sporophores may be formed inside specially designed fruiting body which has a protective wall of fungal tissue *ex.* Uredinales and Ustilaginales.

4.3 Pycnidium

In certain group of imperfect fungi the conidia arise in globose or flask shaped bodies known as pycnidia *ex. Guignardia.* The conidiophores arise from the internal cells of the pycnidial wall. The pycnidial wall is pseudoparenchymatous. Pycnidia may be completely closed or may have an opening, the ostiole. The pycnidium may be provided with a small papilla or with a long neck leading to the opening; they vary greatly in size, shape, color and consistency of wall. They may be superficial or sunk in the substratum; they may be uniloculate, simple or labrinthiform. A fruit body similar to pycnidium is formed in sexual cycle of the rust fungi (Uredinales). They are known as pycnia or spermogonia and produce pycniospores or spermatia.

4.4 Aecia

An aecium is a group of binucleate hyphal cells within the parasitized host, which give rise to spore chains by successive and conjugate division of the nuclei. These are mostly shallow or deep cup shaped structures, with or without peridium, and have the sporophore packed in a layer at the bottom of the cup. As the aecium develops, the disjuncture cells disintegrate and the spores thus separate from each other.

4.5 Ascocarp

Ascomycetes produce their asci in fruiting bodies called ascocarp. Ascocarp contains asci in which usually definite number (8) of ascospores are formed.In these spores the chromosome number is half. Simple Ascomycetes bear naked asci without fruiting body. In general there are 5 major categories, separated according to way they bear their asci.

4.5.1 Gymnothecium

In *Gymnoascus* there is a loose open network of peridial hypae. The asci can be seen through the network. In *Myxotrichum* certain peridial hypae extend as hooked hair.

4.5.2 Cleistothecium

Those which produce their asci inside a completely closed globose, broadly oval or pear shaped ascocarp. It has a definite wall of fungal hyphae. Asci inside the ascocarp are found usually scattered *ex. Eurotium* or arising in tufts from the basal region of ascocarp *ex. Erysiphe.* The asci released by irregular disintegration of the wall of cleistothecium.

4.5.3 Perithecium

This is flask shaped structure, opening by a pore or ostiole. There is a well defined wall, which is made up of sterile cells derived from hyphae, which surrounded the ascogonium during development. The perithecia are often single *ex. Sordaria* and *Neurospora*, but in some genera *ex. Xylaria* and *Hypoxylon*, they are embedded in or seated on a mass of tissue forming a perithecial stroma. Asci are released through a pore. The asci are generally unitunicate. The asci arise from a basal plectenchymacentrum. Asci are usually clavate or cylindric in shape with a distinct stipe. The pore may be lined with hair like periphysis. From the inner wall of perithecium, a layer of asci develops. Sterile paraphyses may arise from the base of the perithecium *ex. Neurospora.*

4.5.4 Apothecium

Those which produce their asci in a saucer shaped open ascocarp. The asci present in a layer called hymenium. The asci are freely exposed at maturity *ex. Peziza.*

4.5.5 Ascostroma

Composed of vegetative hyphae modified into pseudoparenchyma. The fruit body lacks its own distinct wall. In most cases the asci are bitunicate. In ascostroma several levels are present. The asci are contained in one or several cavities (locules) formed within a pre-existing ascostroma *Ex. Guignardia.*

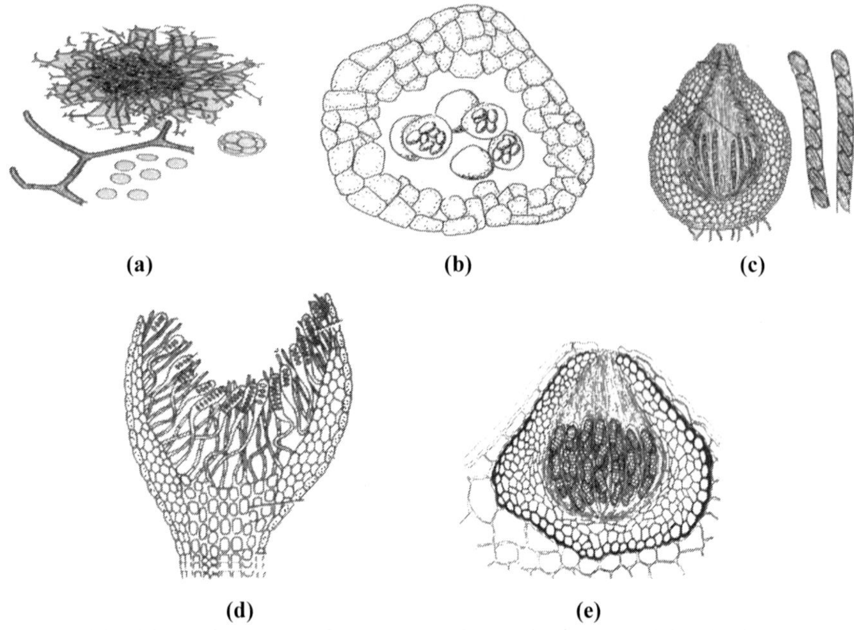

Fig. 4.3. Various types of ascocarp **(a)** Gymnothecium **(b)** Cleistothecium **(c)** Perithecium **(d)** Apothecium **(e)** Ascostroma

4.6 Basidiocarp

The higher basidiomycetes produce their basidia in highly organized basidium (spore mother cell). The basidium is a club shaped structure bearing on its surface a definite number of basidiospores (typically four) which are usually formed as a result of karyogamy and meiosis. It is separated from the rest of the hypha by a septum over which a clamp connection is generally found. The fruiting body is known as basidiocarp, which is a compound sporophore. The sexual spore, basidiospores are born exogenously on basidiocarp. Basidiospores are typically a unicellular, uninucleate haploid structures. Basidiocarp may be thin and crust like, gelatinous, cartilaginous, papery, fleshy, spongy, corky, woody or of any texture. They very greatly in size, from microscopic to 3 feet or more in diameter. Most Basidiomycetes bear their basidia in basidiocarps, however in rust and smuts belonging to order Uredinales and Ustilaginales, do not form any basidiocarp except in one or two species.

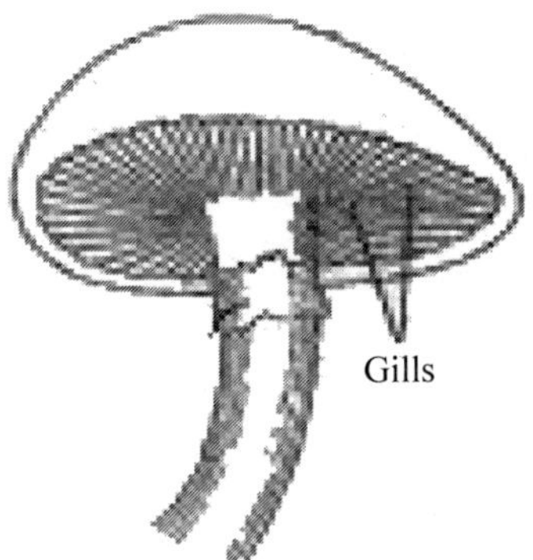

Fig. 4.4. Basidiocarp of *Agraricus*

4.7 Asexual Reproduction

It does not involve the union of nuclei, sex cells (gametes) or sex organs (gametangia). Any asxual spores formed by mitosis are termed as mitospores. Asexual stage is also known as imperfect stage or anamorph stage.These asexual spores are produced numerous in number, since asexual cycle is usually repeated several times in a year. Following are the different methods of asexual repreoduction in fungi.

4.7.1 Fragentation

This is the most common method of asexual reproduction. In this method, the hypha breaks into component cells, which lead to independent life as spores. Arthrospores are spores formed by fragmentation of mycelium *ex. Oidium, Geotrichum*, (Arthros = joint, spore = seed). Sometimes, the content of the cell gets rounded off and surrounded by the wall around it, to escape unfavourable conditions. These cells swell before breaking or separate from each other. Such component cells are known as chlamydospores (chlamy = mantle).

4.7.2 Fission

First the cells of the fungus elongate. The nucleus undergoes mitotic division. The 2 nucleus migrate towards the poles. Transeverse septum develops in the middle of the cell and separates it into 2 cells (daughter cells) which lead an independent life. It is commonly seen in unicellular yeast (*Schizosaccharomyces*).

4.7.3 Budding

In this method first an outgrowth bud is formed from parent cell. Meanwhile the nucleus of parent cell under goes mitotic division. Then one of the daughter nucleus along with the protoplasm moves into the bud and a constriction develops, bud enlarges and eventually get detached from the parent cell. Spores produced after budding are known as blastospores ex. Yeast. Sometimes multiple buds are also seen i.e., bud over bud. This type of budding is seen in yeast (*Saccharomyces cereviseae*).

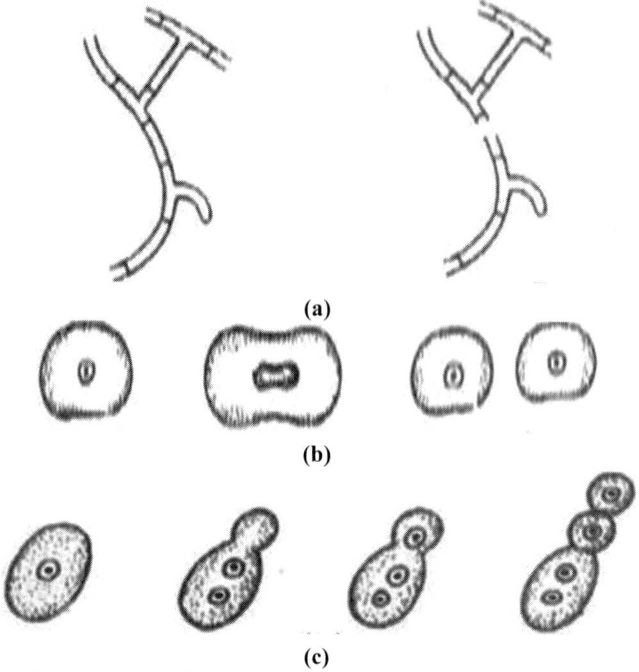

Fig. 4.5. Types of vegetative reproduction in fungi **(a)** A mycelial fragmentation **(b)** Stages of fisson in yeast **(c)** Stages of budding in yeast

4.7.4 Sporulation

Sporulation is the process of the production of spores. A spore is a minute propagating unit, function as a seed, but, differs from it in lacking a preformed embryo. The spores are of two types: sporangiospores and conidia.

4.7.4.1 Sporangiospores

When the spores are produced internally i.e. within the sporangia, are called sporangiospores. These are asexual spores. The sac like structure which produced on sporangiophores is known as sporangium (spore = seed, angious = vessel). In sporangia the entire protoplasmic contents are converted, through cleavage into one or more spores. The special type of hyphae, which bears sporangia is called sporangiophore. Sporangiophores may or may not be distinguishable from hypha. It may be simple or branched. Sporangia may vary in size, shape, production and number of spores.

4.7.4.1.1 Zoospores (Zooanimal, sporos = seed, spore)

Zoospores are characteristic of the Mastigomycotina. They usually lack cell walls. The cytoplasm remain enveloped in a membrane derived from the vacuoles and the plasmalemma. The zoospores produce flagella either during or after cleavage of cytoplasm. The sporangium which produces zoospores are known as zoosporangium. Flagella are originated from a granular structure, blipharoplast in the cytoplasm. Based upon the number, position and nature of flagella, following- types of zoospores are as follows, zoospores with single posterior whiplash flagellum *ex. Allomyces*, *Blastocladiella*, biflagellate zoospores with two whiplash flagella of unequal length (anisokont) ex. Myxomycotina and Plasmodiophoromycetes, zoospores with single anterior tinsel flagellum *ex.* Hypoditridiomycetes biflagellate zoospores with one flagellum of whiplash type and one of the tinsel type *ex. Phytophthora* (heterokont). Flagella are originated from a granular structure, blepharoplast in the cytoplasm Flagella of zoospores are of two types:

4.7.4.1.1.1 Whiplash Flagella

In this type, the basal portion is thick and rigid, while the upper portion is short, narrow and flexible. It gives a whip like appearance to flagellum (lower or basal portion is much longer than the upper or terminal portion). Each whiplash flagellum has 11 microtubules arranged in 9+2 pattern. The microtubules enclosed in a smooth membranous axoneme sheath continuous with the plasma membrane.

4.7.4.1.1.2 Tinsel Flagella

It is a feathery structure consisting of long rachis with lateral hair like structures known as Mastigonemes or flimmers on all sides along its entire length. The nature, number and position of flagella play an important role in classification of fungi. Uniflagellate zoospores possess one flagellum and biflagellate zoospores possess two flagella. The two flagella may be equal or unequal (heterokont).

4.7.4.1.2 Monosporous sporangiola

In this sporangia only one sporangiospore is produced inside a sporangium *eg. Choanephora*. In general sporangium contains numerous spores *ex. Mucor, Rhizopus*. Columella are vesicle or central sterile region continuous with sporangiophore inside sporangia *ex. Rhizopus*

4.7.4.1.3 Merosporangia: Spores borne linearly within a sporangial wall, such sporangia lack columella *ex. Syncephalastrum*

4.7.4.1.4 Sporangiola: Few spored sporangium are known as sporangiola *ex. Blakeslea.*

Sporangiospores which are produced inside the sporangia are of two types: aplanospores and zoospores (Planospores).

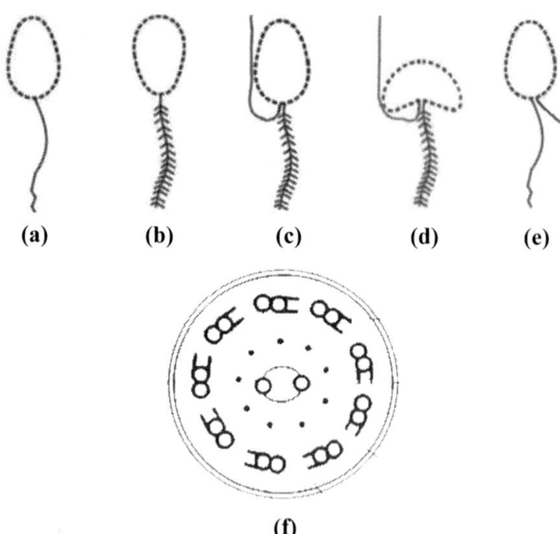

Fig.-4.6: Different kinds of zoospores in different group of fungi **(a)** Chytridiomycetes **(b)** Hypochytridiomycetes **(c)** *Saprolegnia* **(d)** Oomycetes **(e)** Myxomycetes and Plasmodiophoromycetes **(f)** Transverse section of flagellium to show 9 + 2 arrangement

4.7.1.5 Aplanospores (A = not + planetes wanderer sporos = spore)

Sporangiospores without flagella are known as aplanospores. These are non motile *ex. Rhizopus, Mucor*. Sporangiospores are asexual spores packed internally in globose or cylindrical sporangia, borne on sporangiophore. Sporangiophore may or may not be distinguished from hypha. The spores may be uni- or multi-nucleate and are unicellular, thin, smooth walled and globose or ellipsoid in shape. They are formed by cleavage of the sporangial cytoplasm. The special type of hyphae which bears sporangia is known as sporangiophore. In *Mucor* and *Rhizopus* columella is present, which is a vesicle or central sterile region continuous with sporangiophore inside sporangium.

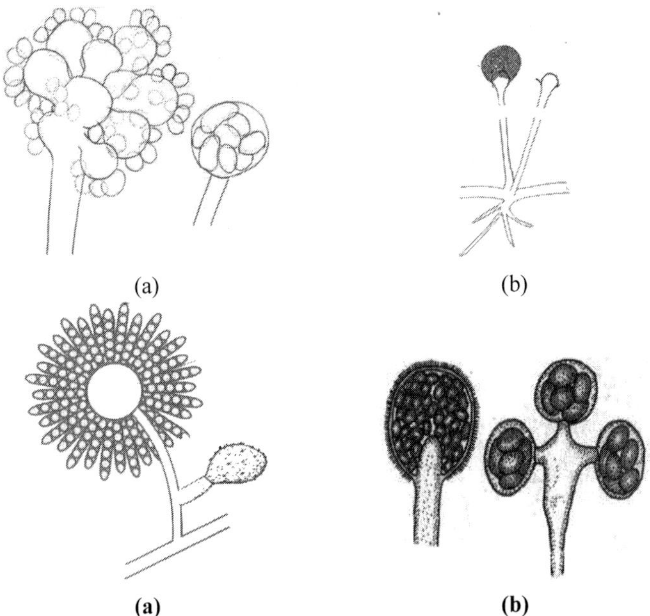

Fig. 4.7. Sporangia in Mucorales **(a)** Monosporous sporangiola of *Choanephora* **(b)** Sporangia of *Rhizopus* **(c)** Merosporangia of *Syncephalastrum* **(d)** Sporangiola of *Blakeslea*

4.7.4.2 Conidiospores

These are asexual reproductive structure. Conidia are produced exogenously on conidiophores which may be simple or branched. The cells which produce conidia are known as conidiogenous cells.

In some cases, they are produced in specialized structures which have conidiogenous cells ex. pycnidia and Acervuli. Two basic types of conidial development and conidia are present in fungi, thallic (thallospores) and blastic (blastospores).

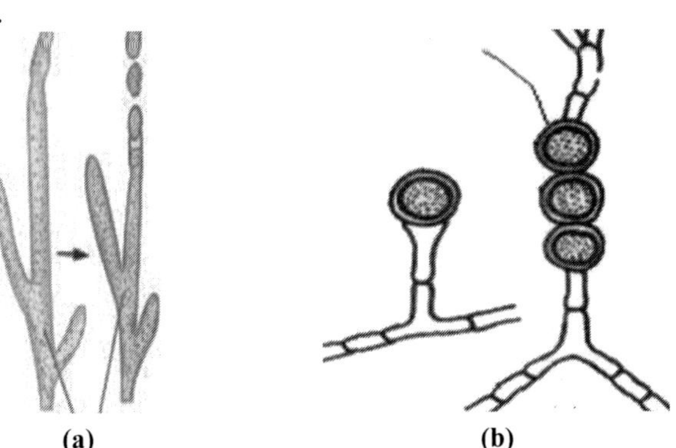

Fig. 4.8. Two types of thallospore **(a)** Arthrospores **(b)** Terminal and intercalary chlamydospores

4.7.4.2.1 Thallic (Thallospores)

There is no enlargement of the conidial initial i.e. the conidium are developed by conversion of a pre-existing segment of the fungus thallus. Thallospores are set free by decay or disarticulation of the parent hypha *ex. Geotrichum.* Two types of thallospores are present; arthrospore and chlamydospores. Arthrospore arise by close septation in basipetal succession. Each cell round off and set free as thin walled arthrospores *ex. Oidum.* Chlamydospore may be produced when the protoplasmic content of a short length hypha accumulate at a particular point, round off and surrounded by a thick often pigmented wall. They are formed terminally or intercalary in a hypha. They are means of asexual survival structure especially in soil fungi *ex. Fusarium* and *Pythium.* They have dense content and rich food reserves, oil or glycogen. Old hyphae of water mould, *Saprolegnia* produce chlamydospores single or in chains. They may be disperse in water current and then known as gemmae. Thick walled dikaryotic spores of smut fungi are also known as chlamydospores.

4.7.4.2.2 Blastic conidia: (Blastospores or true conidia)

In blastospores, there is enlargement of the conidium initial before it is delimited by a septum and the conidium develop from part of a cell. This type of development is known as blastic. Two main kinds of blastic development have been distinguished, holoblastic and enteroblastic.

4.7.4.2.2.1 Holoblastic: In this type, both the inner and outer wall layers of the conidogenous cells contribute to conidium formation *ex. Pleospora herbarum.*

4.7.4.2.2.2 Enteroblastic: In this type of deveploment, only the inner wall or completely new wall layer involved in conidium formation. This development may be threatic (porospore) and phialidic.

In Tretic conidia: the inner wall layer balloons out through a narrow pore or channel in the outer wall layer *ex. Helminthosporium* and *Alternaria.*

Phialidic (*Phialospores*): The conidogenous cell is a specialized cell, phialide. During the formation of first conidium, the tip of the phialide is ruptured and further conidia develop.

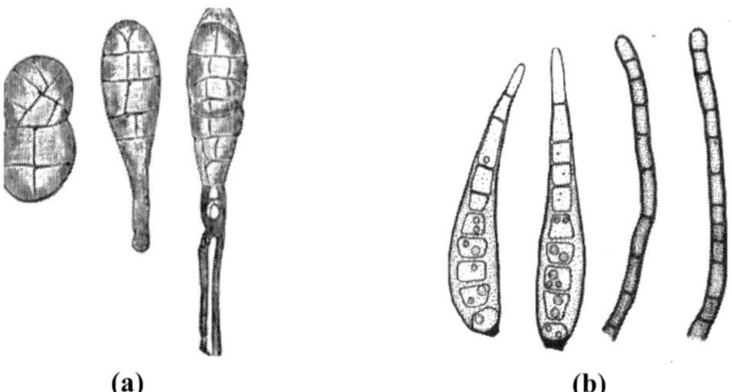

Fig. 4.9(a) Blastic conidia **(a)** Holoblastic **(b)** Enteroblastic tretic

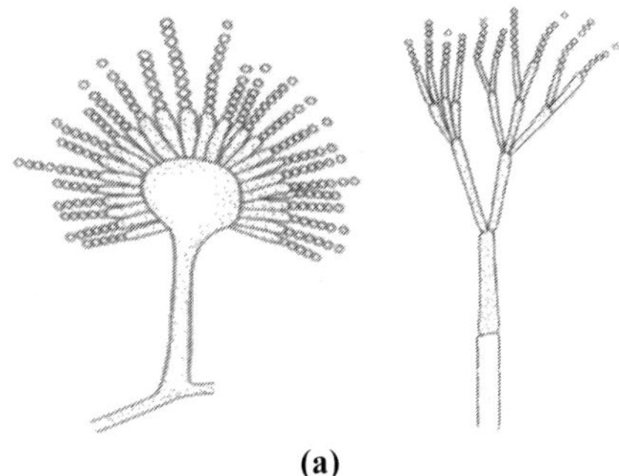

(a)

Fig. 4.9(b) **(a)** Enteroblastic phialidic

4.8 Sexual Reproduction

Sexual reproduction involves the union of two sex nuclei, two sex cells, or two sex organs or two somatic cells or hyphae for the formation of new individuals. The two sex nuclei unite to form new sexual spores. Sexual stage is also known as perfect stage. These are produced in a definite number and are thick walled and resistant to unfavourable conditions. Sexual spores play an important role in perpetuation or spreading of fungus from one season to another season.

4.9 Sex Organs of Fungus

Sex organs of fungus are known as gametangia and sex cells are known as gametes. (gamete = husband, angion = vessel). Gametangia may contain one or more gametes or gamete nucleus.When gametangia do not differ morphologically, they are known as isogametangia. When gametes are morphologically similar they are known as isogametes. In heterogametangia the gametangia differ morphologically.When gametes differ morphologically they are known as heterogametes. Male gametangium is known as antheridium, which is generally club shaped. Female gametes are known as egg or oosphere. Motile gametes are known as planogametes. Male and female gametes are differentiated symbiotically by using + or – signs respectively.

4.10 Phases of Sexual Reproduction

In sexual reproduction the fungi involves union of two compatible nuclei. The sexual reproduction consists of 3 phases; plasmogamy, karyogamy and meiosis.

4.10.1 Plasmogamy

Plasmogamy (Plasma = amolded object, Gamos = marriage) bring two haploid nucleus togather in one cell. In this phase union of two protoplast takes place.As a result of it, the two nuclei come together within the same cell. The protoplasts of antheridia are transferred into Oogonia. Male and female nuclei lay together.

4.10.2 Karyogamy

The fusion of two nuclei brought togather by plasmogamy is known as karyogamy (Karyon = nucleus, gamos = marriage). It results in the formation of diploid nucleus (zygote). In case of lower fungi karyogamy follows plasmogamy almost immediately. But in higher fungi, karyogamy is delayed. As a result the nuclei will remain in pairs. A binucleate cell containing one nucleus from each parent is formed. Such a pair of nuclei is known as dikaryon and the process is known as dikaryogamy.

4.10.3 Meiosis (meiosis- reduction)

In this phase reduction division occurs, which reduces the number of chromosomes to the haploid. Diploid nucleus results in 4 haploid nuclei.

In a true sexual cycle, the above said three processes occur in a regular sequence and usually at specific points in the life cycle. Fungi differ from each other in respect of the methods, by which the two opposite nuclei are brought together in one cell.

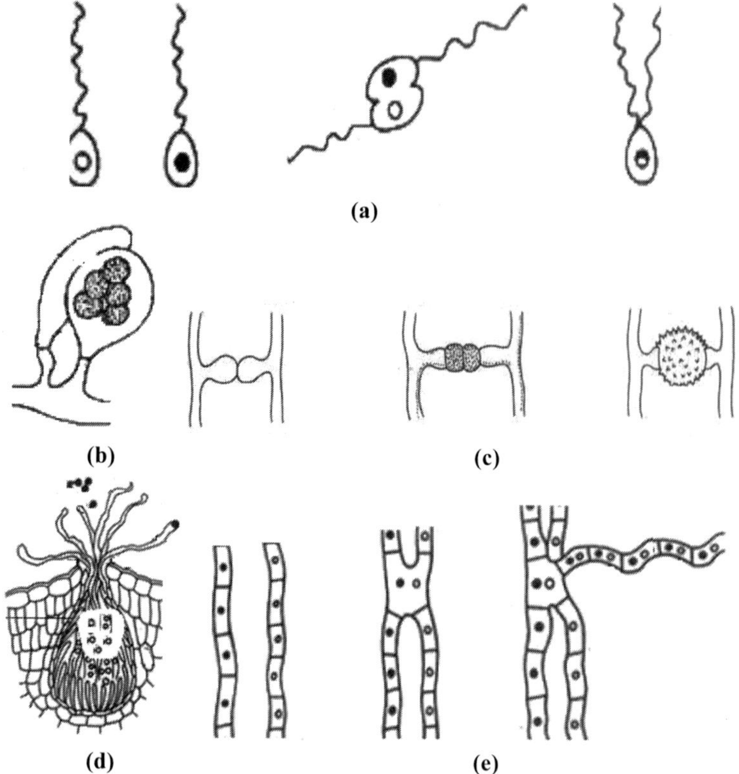

Fig. 4.10 Different methods of Plasmogamy in fungi **(a)** Planogamatic copulation **(b)** Gametangial contact **(c)** Gametangial copulation **(d)** Spermatization **(e)** Somatogamy

4.11 Methods of Sexual Reproduction

Plasmogamy takes place by following methods, planogametic copulation (gametogamy), gametangial contact (oogamy), gametangial copulation, (gametangiogamy), spermatization and somatogamy.

4.11.1 Planogametic copulation

It involves the union of two naked gametes, one or both of which are motile. After plasmogamy, gametes lose their identity.This method is common in lower fungi. *ex. Synchytrium endobioticum*. This may be further distinguished as isogamous, anisogamous and heterogamous planogametic copulation.

4.11.1.1 Isogamous copulation: both the gametes are motile and similar in size and morphology *ex. Synchytrium.*

4.11.1.2 Anisogamous copulation: The gametes are molite and morphologically similar but differ in size usually male is smaller and the females larger *ex. Allomyces.*

4.11.1.3 Heterogamous copulation: The female gamete (oogonium or ascogonium) is larger and non- motile and the male is smaller and motile. The male gamete, antherozoids or spermatozoides enter the oogonium and fertilize the egg. *ex. Monoblepharis.*

4.11.2 Gametangial contact

None of the gametes is motile. Here the gametangia of opposite sex came in contact. At the place of contact 'dissolution' of wall occurs and a fertilization tube is formed. Then the contents of male gametangia are passed into female gametangia. Even after plasmogamy gametangia do not lose their identity, *ex. Pythium aphanidermatum* and *Phytophthora* sp.

4.11.3 Gametangial copulation

IIt involves the union of two isogametangia. When they come in contact, the wall at the place of contact dissolves, and the entire contents of two gametangia fuse to form a single unit. Gametangia lose their identity. The two protoplasts mix and the unit increases in size, *ex. Rhizopus stolonifer* and *Mucor* and *yeast.*

4.11.4 Spermatization

Minute uninucleate male cell, known as spermatia, are produced on spermatiophore. They are carried to female structure, receptive hyphae.At the place of contact, the wall dissolve and content of spermatia pass into receptive hypha and dikaryotic nucleus is formed *ex. Puccinia graminis tritici* and *Podospora.*

4.11.5 Somatogamy

Many higher fungi do not produce sex organs. In such cases somatogamy takes place. It is the union of two somatic hypha or somatic cells to form sexual spores. *ex. Agaricus campestris.* This is also known as pseudomixis.

Various terms have been proposed for copulation between two vegetative cells.

Pseudogamy: Copulation between two vegetative cells which are not closely related to each other *ex. Peniophora sambuci.*

Pedogamy: Pseudomictic copulation between mature and immature cells *ex. Yeast.*

Adelphogamy: Pseudomictic copulation between mother and daughter cells, *ex. Zygosaccharomyces chevalieri.*

Automixis: In automixis there may be copulation between 2 cells of the female gametangium *ex.* Higher Ascomycetes.

Parthenogamy: It is the union of two incompatible gametes *i.e.*, fusion of two female gametes. *ex. Ascobolus.*

Autogamy: Fusion of nuclei in pair within a single cell of the female gametangium *ex. Humaria granulate.*

4.12 Parasexualism

It is the process by which genetic recombinations can occur within fungal heterokaryons, by fusion of the two nuclei and formation of a diploid nucleus. In a few mitotic divisions crossing over occur and results in the appearance of genetic recombinants, as the diploid nucleus progressively and rapidly loses individual chromosomes, to revert to its haploid state. Considering that fungi exist and grow primarily, as adjacent hyphae, that may form heterokaryons as a result of anastomoses or fertilization,

The frequency of parasexualism may equal or surpass, that brought about by sexual reproduction.

4.13 Different types of Sexual Spores

4.13.1 Oospore

Oospore is a thick walled spore that develops from an oosphere through either fertilization or parthenogenesis *ex. Monoblepharis.* In lower Mastigomycotina (*ex. Synchytrium*) by isogamous copulation, diploid zygote nucleus (synkaryon) is formed. The resultant diploid zygote cell develops a thick wall around it with food reserve in it. It thus becomes resting spore (in *Synchytrium*) or an oospore in Oomycetes *eg. Pythium* and *Phytophthora.* In Oomycetes oospore development begin with the formation of oosphere within the oogonium. After fertilization union of antheridial nucleus with ooplasm, a thick outer wall is formed around it. This is known as oospore, which is diploid. It contains reserve food in the form of lipid. These are sedentary and important in survival. In *Saprolegnia* spore wall is double layered, the outer epispore is the initial zygotic membrane with the smooth oosphere membrane which contains cellulose.The inner endospore layer is usually thicker and stratified. The oospore germinate by germ tube which may bear a zoosporangium or continue to grow as coenocytic mycelium.

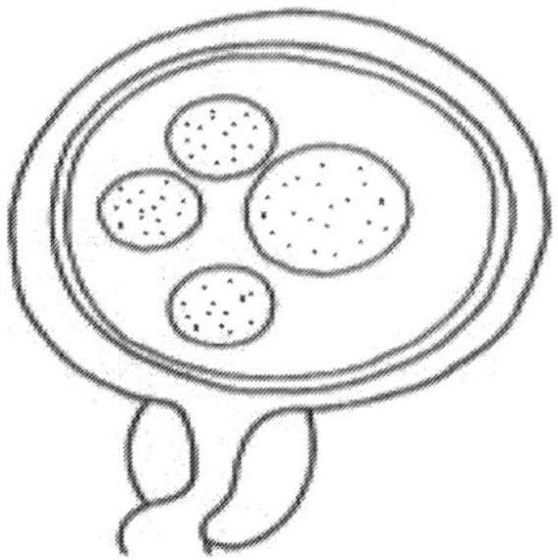

Oogonia with amphignynous antheridia

Fig.4.11. Sexual reproduction in *Phytophthora*

4.13.2 Zygospore

Zygospores are produced by Zygomycetes. These are sexually produced resting structure, formed by fusion of equal or unequal gametangial copulation. Karyogamy may occur early or before meiosis and zygospore germination. The zygospore develop within zygosporangium. These gametangia may arise from same mycelium or from different mycelia.Mature zygospores are often large and thick walled, warty structure with abundant lipid reserves. The zygospore germinate after a period of long rest by a germ tube, or by the formation of germ sporangium. During germination, the outer wall cracks and the inner comes out in the form of germ sporangiophore, which bear a terminal single sporangium containing many spores. Azygospore are similar to zygospore but develop parthenogenetically in zygosporangia, arising from one or both of two gametangia.

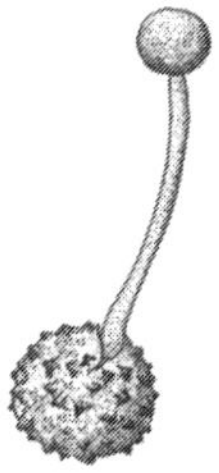

Fig.4.12. Zygospore of *Rhizopus*

4.13.3 Ascospores

Ascosporess are meiospores borne in an ascus in Ascomycetes. They are formed in developing acus as a result of nuclear fusion immediately followed by meiosis. Eight ascospores are typically formed within the ascus by free cell formation. Except Hemiascomycetes, the asci are produced in fructification known as ascocarp. In non-explosive asci, the ascopores are released by breakdown of the ascus wall. However, in most cases the asci are explosive and infect being turgid cell, that eventually burst violently, liberating the contained ascospores. Ascospores vary greatly in size, shape and colour and other characteristis. The ascospore size varies from 4-5 × 1 μm (in *Dasyscyphus*) to 130 × 45 μm (lichen *Pertusaria pertusa*). The shape of the ascospores varies from globose to oval in *Saccharomyces cerevisiae* and *Erysiphe*, broadly.

Lenticular in *Eurotium*, filiform in *Claviceps*, reniform in *Fabospora*, cylindrical in *Endomycopsis selenospora*, lemon shaped, cylindrical, sausage shaped in *Phyllachora* sp., saturn shaped in *Hansenula saturnus*, needle shaped in *Nematospora*, hat shaped in *Hansenula* sp. and *Pichia*, pully wheel in *Emericella* and crescent in *Ceratocystis*. Ascospores may be uninucleate, binucleate (*Neurospora crassa*) or multinucleate, unicellular or multicellular divided by transverse or by transverse and longitudinal septa ex. *Cordyceps*. The ascospore wall may be thin or thick, hyaline (*Peziza*) or coloured (*Xylaria*), smooth or reticulate folds (*Tuber*), warty, spiny or have characteristic longitudinal ribs (*ex. Neurospora*), pits (*Gelasinospora*), appendages (*Bombordia, Podospora*) or have gelatinous sheath (*Sordaria*). In many cases ascospores are resting structures which survive adverse condition. They have food reserve in the form of lipids and sugar (trehalose). Ascospores are important as a mean of dispersal, survival and in genetic recombination.

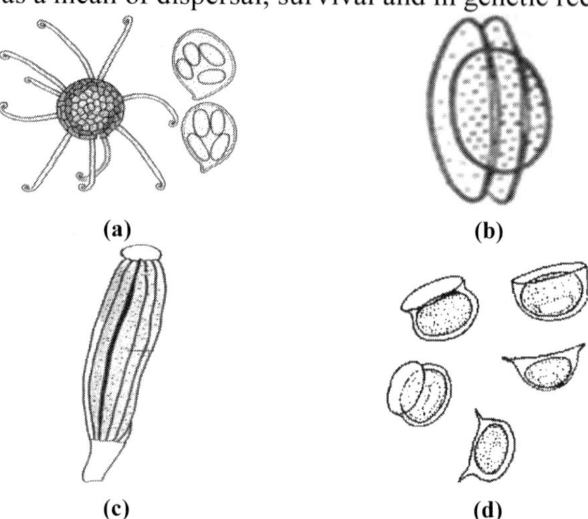

Fig. 4.13. Various types of ascospores **(a)** Globose (*Erysipha*) **(b)** Lenticular (*Emericella*) **(c)** Filiform (*Claviceps*) **(d)** Hat shaped (*Hansenula*)

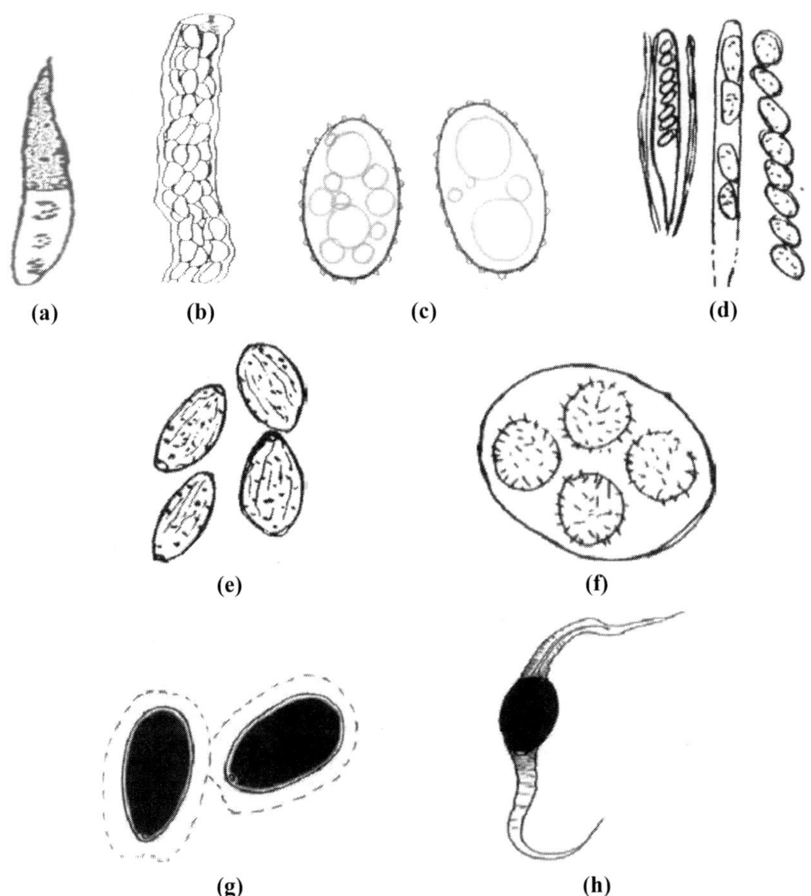

Fig. 4.14. Various types of ascospores **(a)** Needle shaped (*Nematospora*) **(b)** Multiseptal (*Cordycep*) **(c)** Hyaline (*Peziza*) **(d)** Coloured (*Xylaria*) **(e)** Longitudinal ribs (*Neurospora*) **(f)** Warty (*Tuber*) **(g)** With gelatinous sheath (*Sordaria*) **(h)** Appendaged (*Podospora*)

4.13.4 Basidiospores

Basidiospores are the sexual spores and is typically unicellular, haploid structure containing one or two haploid nuclei. Their size varies from 3 to 20 µm. Basidiospores are normally found in groups of four and attached by tapering sterigmata to the cell, which bears them. In shape, basidiospores are asymmetric and vary from globose, oval, elongated, sausage shaped to almond shaped and the wall may be smooth or ornamented with spines, ridge, or folds. The colour of basidioispore is important for identification. They may be colourless, white, cream, yellowish, brown, pink, purple or black. The spore colour may be due to pigment in the spore cytoplasm or in the spore wall.

Generally, basidiospores have a flatter adaxial face and a more curved abaxial face. The point of attachment of the spore to the sterimgata is the hilum, which persist as a scar at the base of a discharged spore. Near hilum a small projection hiler appendix is present. This is involved in the basidiospore discharge, in which a drop of liquid formed on the hilar appendix coalesce with a second blob of liquid. On the spore surface creating a momentum, which leads to acceleration. The spore is projected for a short distance from the basidium. Such spores are known as ballisto spores. Violent discharge of the basidiospores occurs in the majority of basidiomycetes except the Gasteromycetes and Ustilaginales. The basidiospore has a distinct germ pore at the opposite end of hilum *ex Coprinus cinereus* and *Agrocybe acericola* but in other basidiomycetes *ex. Flammulina velutipes* and *Schizophyllum commune*, the basidiospores have no pore. Lipid is the major storage reserve product of the spore.

5

Life Cycle of Fungi

There are great variations in the life cycles of fungi. Raper (1954, 1966b) recognizes seven basic types.

5.1 Asexual cycle

In which sexual reproduction apparently lacks and diploidy is restricted to somatic diploids. The entire groups known as Fungi Imperfecti and numerous other species which clearly belong to various groups of the perfect fungi such as *Penicillium notatum* have this life cycle. Certain benefits of sexuality are provided here in many of the sterile forms by para sexual recombination.

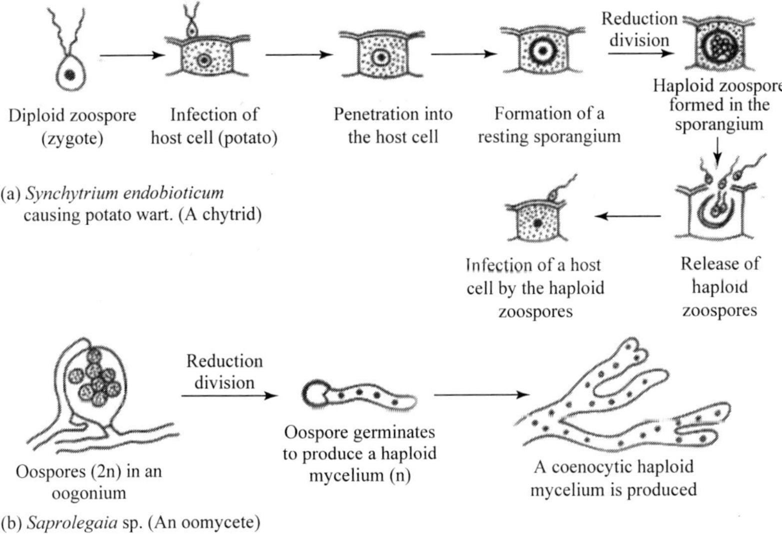

Fig. 5.1. Life cycle of *Saprolegnia* (Ashwathi, 2017)

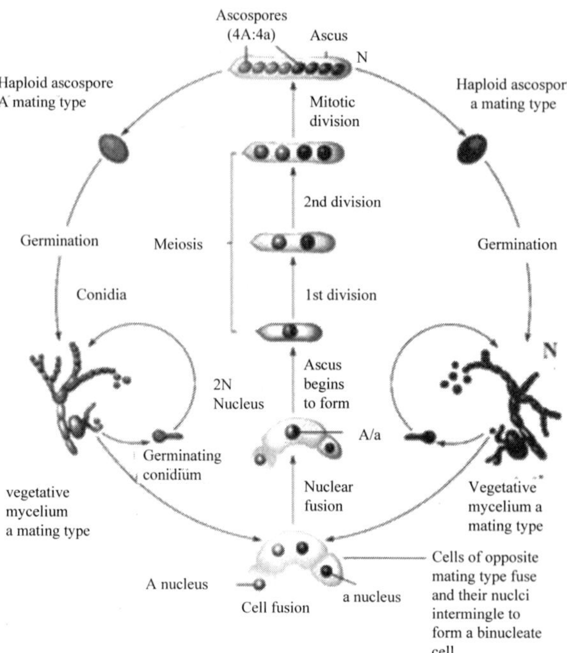

Fig. 5.2. Life cycle of *Neurospora crassa* (Haploid cycle) (Russell, 2010)

5.2 Haploid cycle

This type of life cycle is the simplest possible and is seen in many lower fungi and some Ascomycetes. Here, the diploid phase is confined to the zygote nucleus only.

5.2.1 Haploid cycle with restricted dikaryon

It is similar to the haploid cycle except that paired, potentially conjugant kinds of nuclei lie in close association in some hyphal segment (hence dikaryon) and divide synchronously. Such a life cycle is characteristic of the higher Ascomycetes *e.g., Neurospora.*

5.2.2 Haploid dikaryotic cycle

This cycle differs from the previous one because of the unrestricted and independent growth of the dikaryotic phase. Here, the mycelium derived from germination of a meiospore, may persist in the haploid condition as a monokaryon. But once a dikaryon is formed, it shows potentially unrestricted and independent growth and may comprise the longest phase of the life cycle. This type of life cycle is characteristic of many Basidiomycetes except many of the smut fungi (Ustilaginales).

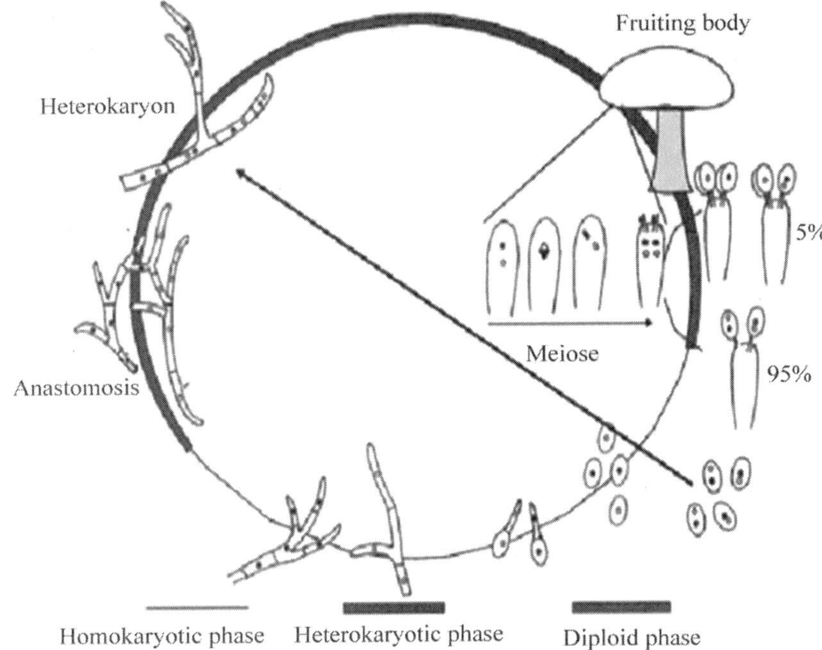

Fig. 5.3. Life cycle of *Agraricus* sp. (*Anton et al.*, 2011)

5.3 Dikaryotic cycle

In this cycle the immediate products of meiosis, ascospores or basidiopores, fuse immediately to reform a dikaryon so that fungus is dikaryotic throughout its life cycle. This type of life cycle is occasionally seen in yeasts and occurs among the smut fungi.

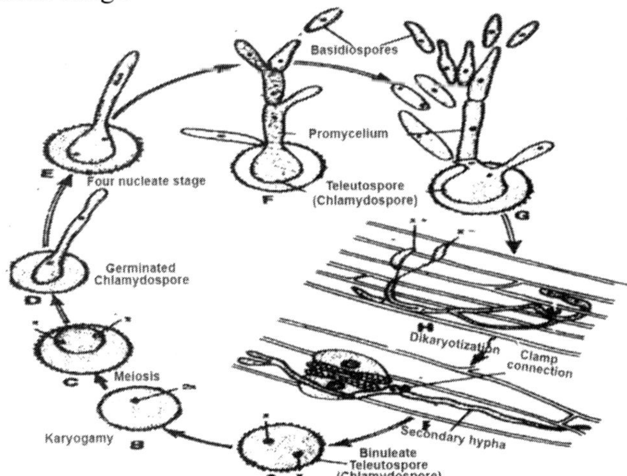

Fig. 5.4. Life cycle of Smut (Dikaryotic cycle) https://biologyboom.com/basidiomycotai

5.3.1 Haploid –Diploid cycle

In this cycle the haploid and diploid phases alternate regularly. Such life cycle is common in algae and higher plants and occurs rarely in fungi and has been described only in two groups. The better known of these is the order Blastocladiales, especially in the section *Eu-Allomyces* of the genus *Allomyces* as well as in *Ascocybe grovesii* (Endomycetales).

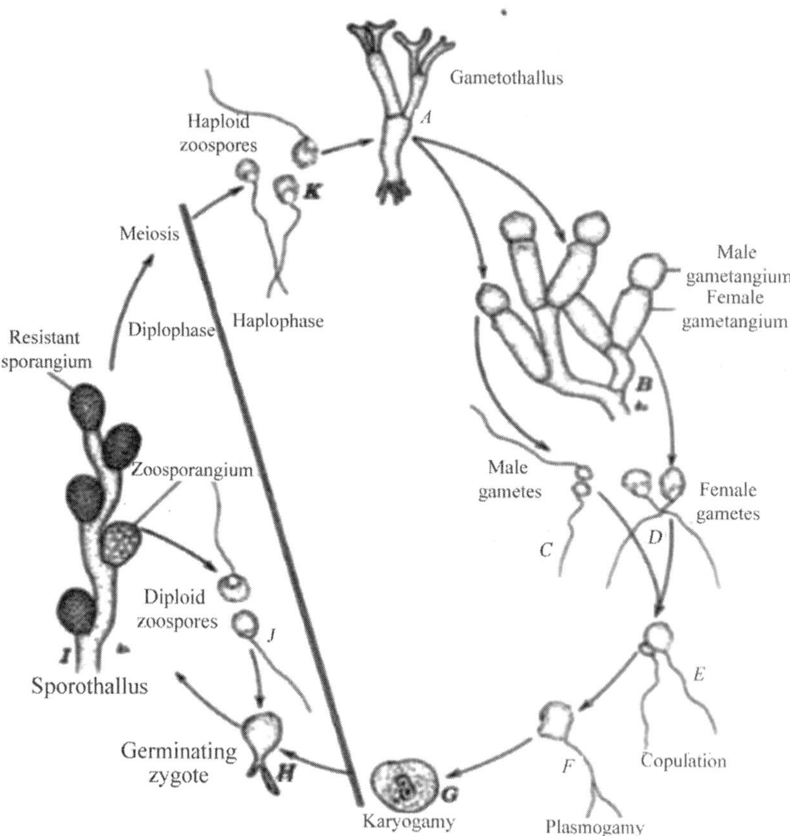

Fig.-5.5. Life cycle of *Allomyces* sp.
(*Source*: Alexopoulos *et al.,* 1996)

5.3.2. Diploid cycle

In diploid cycle, the haploid phase is restricted to the gametes. It seems likely that the majority of oomycetous fungi confirm to this pattern. This type of life cycle is also known to occur in a number of yeasts such as *Saccharomyces ludwigi*, in the Myxomycetes, or true slime molds and also in some members of the Blastocladiales.

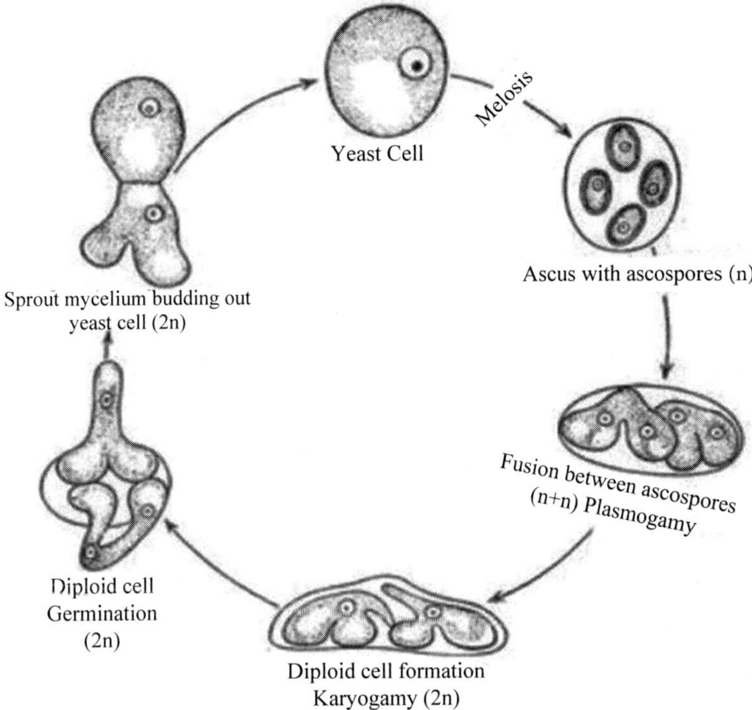

Fig. 5.6. Life cycle of *Saccharomycodes ludwigii* (Diploid).
http://www.biologydiscussion.com

6

Description and Illustration of Selected Fungal Genera

In the classification given below only those groups have been given emphasis which contain plant pathogenic fungi or which are important from taxonomic point of view. The description includes only critical features.

Division	Sub division	Class	Order	Family	Genus
Myxomycota		Myxomycetes	Ceratiomyxales	Ceratiomyxaceae	*Ceratiomyxa*
			Physarales	Physaraceae	*Physarum*
			Liceales	Liceaceae	*Licea*
			Echinosteliales	Echinosteliaceae	*Echinostelium*
			Trichiales	Trichiaceae	*Trichia*
			Stemonitales	Stemonitaceae	*Stemonitis*
			Plasmodiophorales	Plasmodiophoraceae	*Plasmodiophora*
					Spongospora
Eumycota	**Mastigomycotina**	**Chytridiomycetes**	Chytridiales	Olpidiaceae	*Olpidium, Rozella*
				Synchytriaceae	*Synchytrium*
				Phlyctidiaceae	*Rhizophydium*
				Rhizidiaceae	*Polyphagus, Rhizophlyctis*
				Cladochytridiaceae	*Cladochytrium*
				Physodermataceae	*Physoderma*
				Chytridiaceae	*Chytridium,*
				Megachytridiaceae	*Nowakowskiella*
			Harpochytriales	Harpochytriaceae	*Harpochytrium*
			Blastocladiales	Coelomomycetaceae	*Coelomomyces*
				Catenariaceae	*Catenaria*
				Blastocladiceae	*Allomyces Blastocladiella*
			Monoblepharidiates	Monoblepharidiaceae	*Monoblepharis*
			Monoblepharidiates	Monoblepharidiaceae	*Monoblepharis*

Division	Sub division	Class	Order	Family	Genus
				Gonopodyaceae	*Monoblepharella, Gonopodya*
		Hyphochytridiomycetes	Hypochytridiales	Anisolpidiaceae	*Anisolpidium*
				Rhiziomycetaceae	*Rhizidiomyces*
				Hypochytriaceae	*Hypochytrium*
		Oomycetes	**Saprolegniales**	Saprolegniaceae	*Saprolegnia, Achlya*
			Lagenidiales	Lagenidiaceae	*Lagenidium*
			Leptomitales	Leptomitaceae	*Leptomitus*
			Peronosporales	Pythiaceae	*Pythium, Phytophthora, Sclerophthora*
				Peronosporaceae	*Peronospora, Plasmopara Sclerospora, Bremia Basidiophora*
				Albuginaceae	*Albugo*
	Zygomycotina	**Zygomycetes**	**Mucorales**	Mucoraceae	*Mucor, Rhizopus, Absidia Zygorhynchus*
				Pilobolaceae	*Pilobolus, Pilaira*
				Mortierellaceae	*Mortierella*
				Saksenaeaceae	*Saksenaea*
				Endogonaceae	*Endogone*
				Radiomycetaceae	*Radiomyces*
				Thamnidiaceae	*Thamnidium*
				Choanephoraceae	*Choanephora Blakeslea*
				Cunninghamellaceae	*Cunninghamella*
				Helicocephalidaceae	*Helicocephallum*
				Syncephalastraceae	*Syncephalastrum*
				Piptocephalidaceae	*Piptocephalis, Syncephalis*
				Dimargaritaceae	*Dimargaris, Dispira*
				Kickxellaceae	*Kickxella, Coemansia*
			Entomophthorales	Entomophthroceae	*Entomophthora, Basidiobolus*
			Zoopagales	Zoopagaceae	*Zoopage, Stylopage.*
				Cochlonemaceae	*Cochlonema*
		Trichomycetes	Harpellales	Harpellaceae	*Harpella*
			Aselllariales	Asellariaceae	*Asellaria*
			Eccrinales	Palavasciaceae	*Palavascia*
			Amoebidiales	Amoebidiaceae	*Amoebidium*
	Ascomycotina	**Hemiascomycetes**	**Protomycetales**	Protomycetaceae	*Protomyces*

(Contd.)

Division	Sub division	Class	Order	Family	Genus
			Endomycetales	Ascoideaceae	*Ascoidea Dipodascus*
				Spermophthoraceae	*Spermophthora*
				Endomycetaceae	*Endomyces*
				Saccharomycetaceae	*Saccharomyces*
			Taphrinales	Taphrinaceae	*Taphrina*
		Plectomycetes	**Gymnoascales**	Onygenaceae	*Onygena*
			Eurotiales	Eurotiaceae	*Eurotium Elericella*
		Pyrenomycetes	**Erysiphales**	Erysiphaceaae	*Erysiphe, Phyllactinia Uncinula, Sphaerotheca Microsphaera, Podosphaera Oidium, Ovulariopsis*
			Meliolales	Meliolaceae	*Meliola*
			Sphaeriales	Chaetomiaceae	*Chaetomium*
				Xylariaceae	*Xylaria*
				Phyllachoraceae	*Phyllachora*
				Sordariaceae	*Sordaria, Neurospora Podospora*
				Polystigataceae	*Glomerella*
			Clavicipitales	Clavicipitaceae	*Claviceps, Cordyceps*
			Coronophorales	Coronophoraceae	*Coronophora*
			Hypocreales	Hypocreaceae	*Nectria, Hypocrea, Gibberella*
		Discomycetes	**Medeolariales**	Medeolariaceae	*Medeolaria*
			Cyttariales	Cyttariaceae	*Cyttaria*
			Tuberales	Tuberaceae	*Tuber*
			Pezizales	Pezizaceae	*Peziza*
				Morchellaceae	*Morchella, Verpa*
			Helotiales	Sclerotiniaceae	*Sclerotinia*
			Rhytismatales	Rhytismataceae	*Rhytisma*
			Myragiales	Myriangiaceae	*Elsinoe, Myriangium*
				Dothideaceae	*Mycosphaerella*
			Pleosporales	Venturiaceae	*Venturia*
				Pleosporaceae	*Pleospora, Cochliobolus*
			Hemisphaerales	Microthyriaceae	*Microthyrium*
	Basidiomycotina	**Teliomycetes**	**Uredinales**	Pucciniaceae	*Puccinia, Gymnosporangium Uromyces, Phragmidium, Ravenalia, Hemileia*
				Melampsoraceae	*Melampsora, Cronartium*

Division	Sub division	Class	Order	Family	Genus
			Ustilaginales	Graphiolaceae	*Graphiola*
				Ustilaginaceae	*Ustilago, Sphacelotheca Tolyposporium, Sorosporium*
				Tilletiaeceae	*Tilletia, Neovossia, Urocystis Entyloma*
		Hymenomycetes	**Tremellales**	Tremellaceae	*Tremella, Exidia*
			Auriculariales	Auriculariaceae	*Helicobasidium, Auricularia*
			Exobasidiales	Exobasidiaceae	*Exobasidium*
			Dacrymycetales	Dacrymycetaceae	*Dacrymyces*
			Aphallophorales	Cantharellaceae	*Cantharellus*
				Schizophyllaceae	*Schizophyllum*
				Clavariaceae	*Clavaria*
				Hydnaceae	*Hydnum*
				Ganodermataceae	*Ganoderma*
				Polyporaceae	*Polyporus*
				Corticiaceae	
				Cyphelliaceae	
				Hericiaceae	*Hericium*
			Agaricales		
				Tricholomataceae	*Armillaria, Pleurotus*
				Agaricalceae	*Agaricus, Amanita*
				Pluteaceae	*Volvariella*
				Strophariaceae	*Psilocybe*
				Copriniaceae	*Coprinus*
				Cortinariaceae	*Cortinarius*
		Gasteromycetes	**Sclerodermatales**	Sclerodermataceae	*Scleroderma*
			Lycoperdales	Geastraceae	*Geastrum*
				Lycoperdaceae	*Lycoperdon*
			Phallales	Phallaceae	*Phallus*
			Nidulariales	Nidulariaceae	*Nidularia, Cyathus*
	Deuteromycotina	**Hyphomycetes**	**Hyphomycetales**	Moniliaceae	*Aspergillus, Penicillium, Gliocladim, Trichoderma, Verticillium, Botrytis, Sporothrix, Geotrichum,*
			Tuberculariales	Tuberculariaceae	*Fusarium, Tubercularia*
			Agonomycetales	Agonomycetaceae	*Rhizoctonia, Sclerotium*

Division	Sub division	Class	Order	Family	Genus
		Coelomycetes	Sphaeropsidiales	Sphareopsidiaceae	*Phyllosticta, Phoma, Phomopsis Macrophomina, Coniothyrium. Chaetomella, Ascochyta, Darluca Diplodia, Botryodiplodia, Septoria*
			Melanconiales	Melanconiaceae	*Colletotrichum, Gloeosporium Pestalotia, Pestalotiopsis Monochaetia, Melanconium*

6.1 Division (1): Myxomycota

6.1.1 Class – Myxomycetes

Ceratiomyxa Schroret

1. Plasmodium, multinucleated.
2. Spores borne externally on the surface of column like sporophore.
3. Spores globose, on germination release a single quadrinucleate amoeboid protoplast.
4. The protoplast through thread phase develops into eight cells that become flagellate and disperse.

Licea Schrad

1. Cosmopolitan, sporangia minute, form protoplasmodia.
2. Capillitium absent but pseudocapillitium with persistent peridium may present.
3. Spore mass colourless.

Echinostelium

1. The sporophoress minute, stalked, columella present.
2. Peridium disappears early during formation of sporophores.
3. Spores colourless, brown or yellow.

Trichia Haller

1. Presence of abundant tubular capillitium with bright coloured spores, and surrounded by persistent peridium.
2. Sporophores lack columella.
3. Presence of short, long and, free elators.
4. Sporangium dehisces irregularly.

Physarum Pers

1. Peridium and capillitium are calcereous.
2. Spore mass always coloured.
3. The sporophores develop from phaneroplasmodia.
4. Capillitium a network of tubules connecting calcereous nodes.

Stemonitis Roth

1. Cosmopolitan on barks of trees, on litter and on dead bamboo sticks in moist, appears as hair - like, black, stalked sporophores with columella.
2. Capillitium network enclosed in membrane.
3. Spores dark violet-brown. The peridium and capillitium, both noncalcareous.

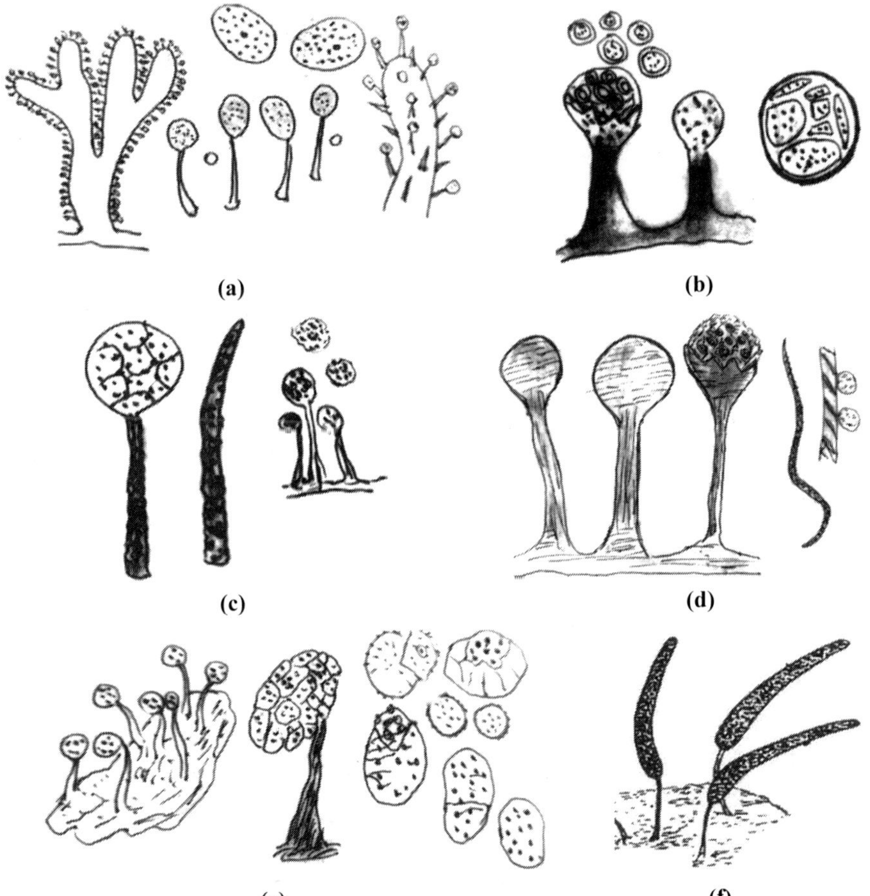

Fig. 6.1. Fruiting bodies of **(a)** *Ceratiomyxa* **(b)** *Licea* **(c)** *Echinostelium* **(d)** *Trichia* **(e)** *Physarum* **(f)** *Stemonitis*

6.1.2 Class – Plasmodiophoromycetes

Plasmodiophora Woronin

1. *P. brassicae* cause club-root disease of cruciferous plants.
2. The root hairs are hypertrophied and grow to form club-shaped swellings on the roots and rootlets.
3. The swollen roots contain by large number of spherical, small resting spores which are released in soil after decay of the roots

Spongospora Brunchorst

1. Presence of cytosori which are sponge-like, ovoid, subspherical or irregulalar and traversed by canals and fissures.
2. *S. subterranea* cause powdery scab of potatoes.

6.2 Division – Eumycota

Subdivision: Mastigomycotina
Class: Chytridiomycetes

Olpidium Braun

1. Thallus holocarpic parasitic in algae, in aquatic fungi, rotifers and on different living and dead materials.
2. The entire content of thallus is converted into zoosporangium.
3. Zoospores very small, tadpole-like, with a spherical head and a trailing flagellum.

Rozella Cornu

1. Parasitic in aquatic fungi and algae.
2. Thallus, holocarpic, endobiotic, develops into zoosporangia or resting sporangia which fills completely into the host cell. *R. allomyces* parasitic in *Allomyces*, *R.blastocladiae* in *Blastocladia.*

Synchytrium de Barry and Woron

1. Causes galls and wart diseases in vascular plants.
2. The thallus is unicellular, naked (wall less).
3. At later stage thallus becomes round in shape with two layered thick, chitinious, golden brown wall (prosorous).
4. Nucleus of prosorus enlarge to a maximum size.
5. Each sporangium with further nuclear divisions forms numerous zoospores. Zoospores and gametes are similar in size and morphology. Uniflagellated zoospores infect host.

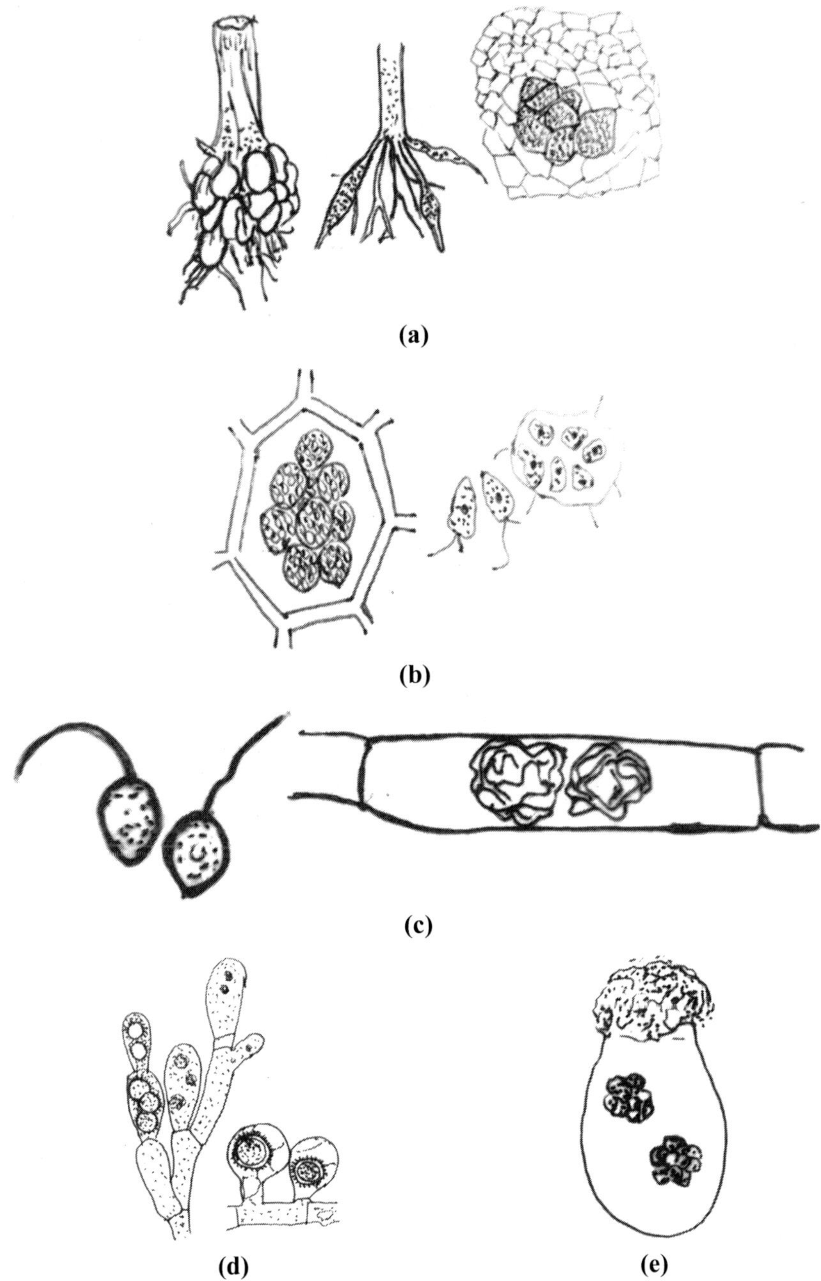

Fig. 6.2. Fruiting bodies of **(a)** Resting sporangia of *Plasmodiophora* in host cell **(b)** Spore balls of *Spongospora* within host cell **(c)** Resting sporangia of *Olpidium* inhost cell **(d)** Resting sporangia of *Rozella* in host cell **(e)** Resting sporangia of *Synchytrium* in host cell

***Rhizophydium* Schenk**

1. The thallus consists of an epibiotic zoosporangium with rhizoids in the substratum.
2. The species grow on pollen, keratin and algae.
3. Uniflagellated zoospores settle on the surface of algae, lose its flagellum, produce rhizoides inside the substratum and act as thallus.

***Polyphagus* Nowak**

1. *P. euglenae* parasitic on encysted *Euglena* and *Chlamydomonas.*
2. Eucarpic and monocentric thallus.
3. Thalli two types, large female thalli and small male thalli.
4. Main body of the thallus behave as prosporangium which develops a sporangium with numerous zoospores.

***Rhizophlyctis* Fischer**

1. On decaying aquatic material, in blue green algae, soil or cellulose rich substrates.
2. *R. rosea* with bright yellow to pink sporangia due to carotenoid pigments.
3. The sporangia with rhizoids.
4. Gametangia absent.

***Cladochytrium* Nowak**

1. On decaying plant parts in water.
2. Thallus polycentric, composed of system of rhizoids with a variable number of sporangia.

***Physoderma* Wallr**

1. *P. maydis* an obligate parasite, causes corn pox or brown spot disease of maize.
2. Resting spores perenating in soil or plant debris, germinate and infect the host.
3. Then finger like endosporangium is formed which shows 20-25 uniflagellate zoospores (resting spores).

***Chytridium* Braun**

1. Parasitic on algae.
2. Thallus monocentric, developing into a sporangium with rhizoidal system.
3. Sporangia pale pink, liberate zoospores through operculum.
4. Resting spores are produced on rhizoidal system within algal host cell.

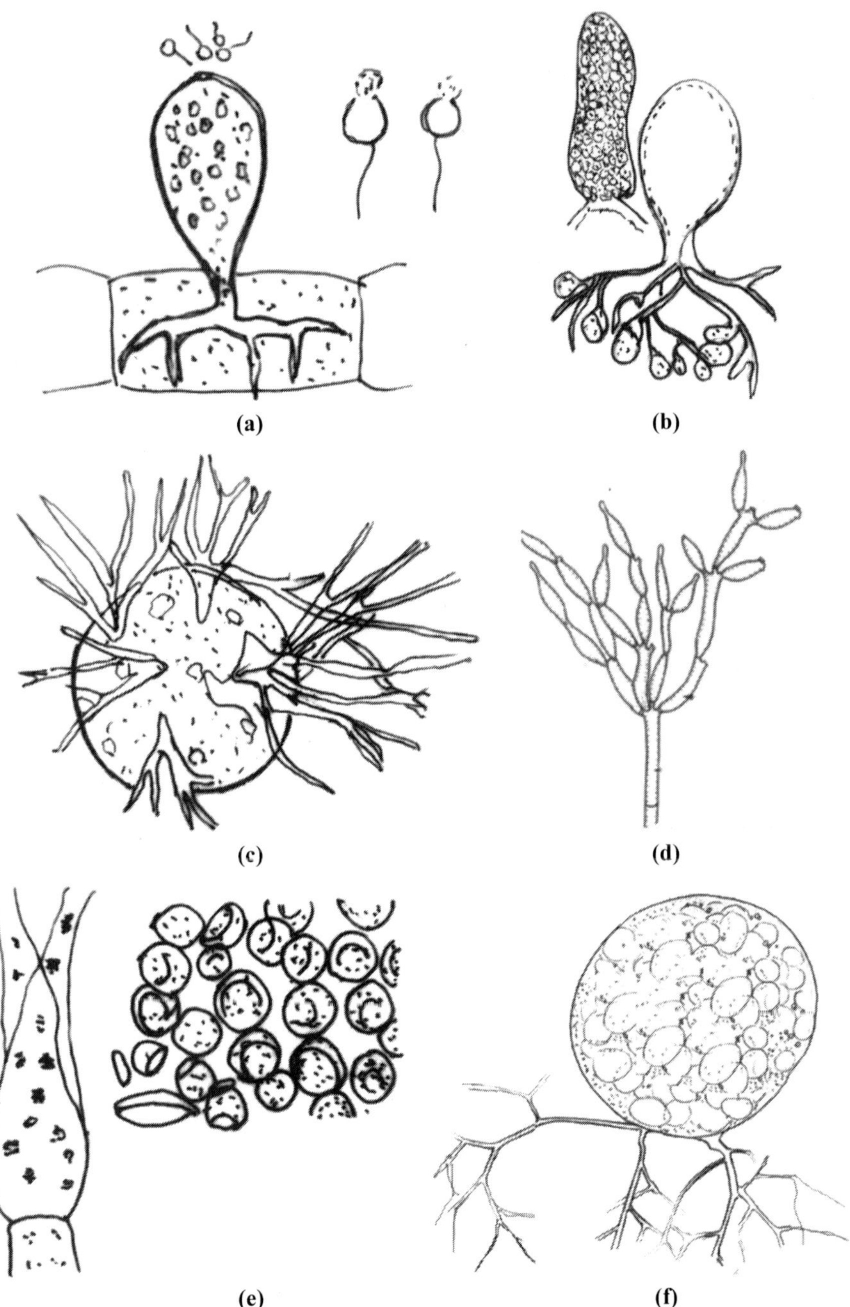

Fig. 6.3. **(a)** *Rhizophydium* **(b)** *Polyphagus* **(c)** *Rhizophlyctis* **(d)** *Cladochytrium* **(e)** *Physoderma* **(f)** *Chytridium*

Nowakowskiella **Schroet**

1. About 12 species, saprobic on plant debris.
2. Thallus profusely branched, filaments occasionally septate, polycentric, bearing terminal or intercalary sporangia, with rhizoid.

Harpochytrium **Lagerh**

1. Thallus fusiform, proliferating, uniaxial, eucarpic
2. Upper sporogenous, lower vegetative, proximal basal disc like holdfast which is superficial, does not penetrate the substrate.

Coelomomyces **Keili**

1. Present in the coelome (body cavity) of mosquito larvae.
2. Thallus non-septate, irregularly branched or lobed hyphae which lack cell wall.
3. Zoospores produced in coelome, fail to reinfect mosquitoes.

Catenaria **Sorokin**

1. Saprophyte or parasite on small insects, nematodes and fungi.
2. Thallus coenocytic, usually unbranched, tubular hyphae swollen at regular distance.
3. Rhizoides present on swollen portions.
4. The swollen portions on septation act as zoosporangia.

Allomyces **Butler**

1. Thallus with a distinct basal parts which has a rhizoidal hold-fast and dichotomously branched pseudoseptate hyphae at the distal end.
2. Zoosporangia and resting sporangia are born on the dichotomous branches.

Blastocladiella **Matthews**

1. Thallus simple, spherical, sac like structure attached to the substratum by means of rhizoids.

Blastocladia **Rensch**

1. The thallus more developed than in Blastocladiella.
2. It consists of basal rhizoidal portion at lower end and two types of sporangia (cylindrical zoosporangia and spherical resting sporangia) at upper end.

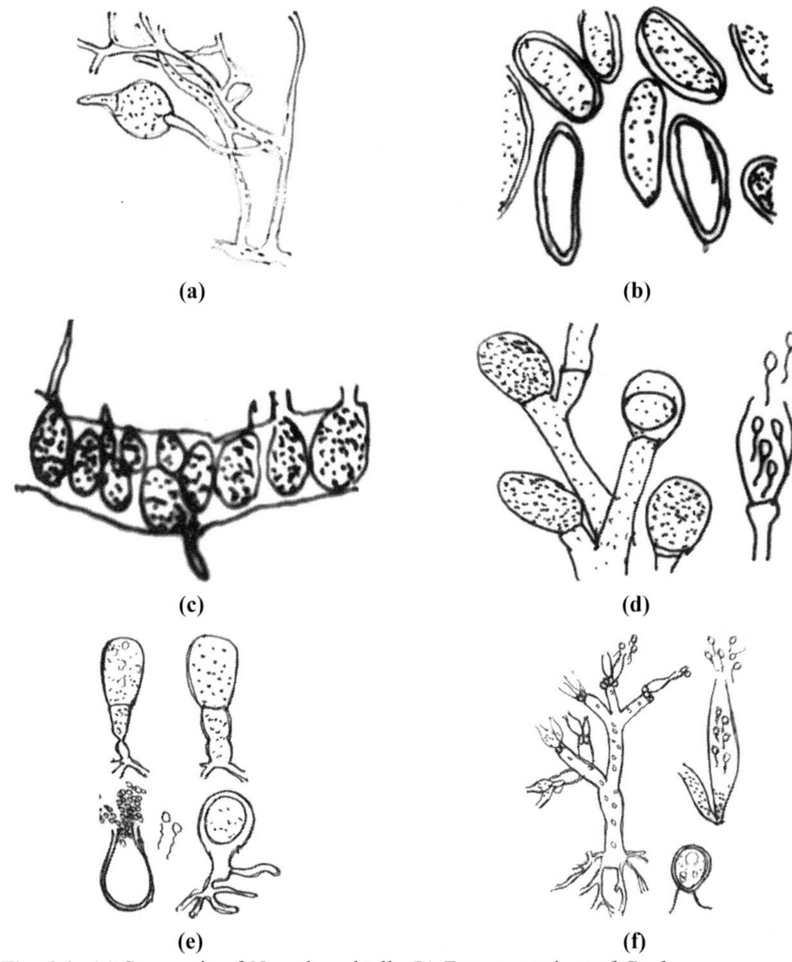

Fig. 6.4. **(a)** Sporangia of *Nowakowskiella* **(b)** Zoosporangium of *Coelomomyces* **(c)** Zoosporangium of *Catenaria* **(d)** Sporangiophore of *Allomyces* **(e)** Zoosporangium of *Blastocladiella* **(f)** *Blastocladia*

Monoblepharis Cornu

1. Grows in still water, attached to the substratum by means of rhizoides.
2. The thallus is like *Allomyces* but without pseudosepta.
3. Cytoplasm is highly vacuolated, with foamy appearance, sporangia elongated, formed at the tip of hypha.

Monoblepharella Sparrow

1. Common in tropical soils, easily isolated on hempseed bait.
2. Mycelium lacks pseudosepta and constriction.

3. Sporangia narrowly ovate, gametangia do not proliferate.
4. Each oogonium bears a typical oospore (rumposome, a tailed body)

Gonapodya Fischer

1. Hyphae with constriction and pseudosepta.
2. Oogonia with several egg-cells.
3. The sporangia are ovoid and produce numerous zoospores

Class – Hyphochytridiomycetes

Anisolpidium Karling

1. 1. Zoosporangia inside host cell with long tube.
2. Zoospore with anteriorly tinsel type flagellum.
3. Sexually formed resting spore (oospore) acts directly as zoosporangium.

Rhizidiomyces Zopf

1. Thallus eucarpic, monocentric, simple, epibiotic
2. Thallus bear a system of branched rhizoids that are sunken into the substratum.

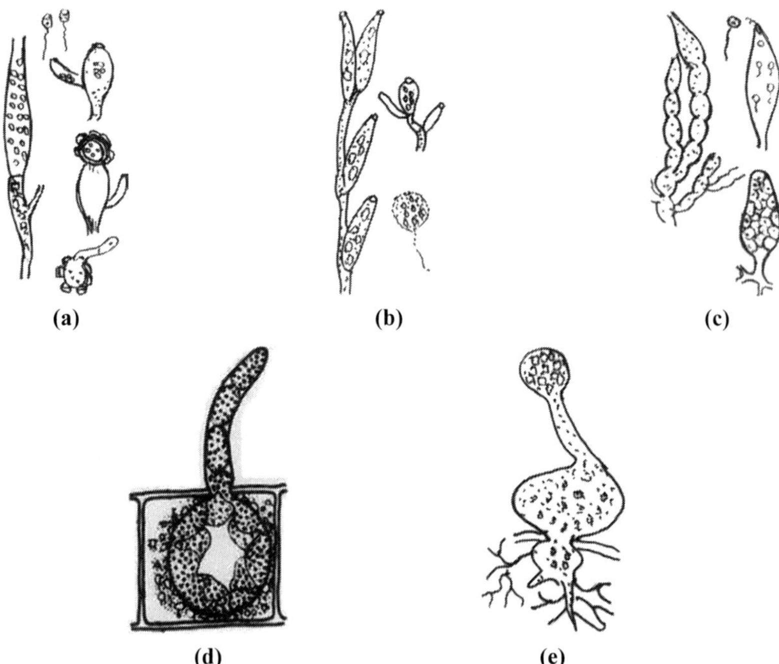

Fig. 6.5 **(a)** Sporangiophore of *Monoblepharis* **(b)** Sporangium of *Monoblepharella* **(c)** Zoosporangium of *Gonopodya* **(d)** Germinating oospore of *Anisolpidium* (e) Zoosporagium of *Rhizidiomyces*

Class-Oomycetes

Saprolegnia Nees

1. Aquatic, on submerged plant or animal debris.
2. Thallus coenocytic, branched hyphae, eucarpic, hyaline, sporangia terminal, elongated, clavate, thicker at the distal end.
3. Proliferation in the sporangia. Zoospores biflagellate.

Achlya Nees

1. Aquatic, saprophytic, rhizoides attached to the substratum.
2. Hyphae broad at base, tapering towards apex.
3. Sporangia, terminal, tubular or clubshaped, separated from vegetative hypha by cross septa.

Lagenidium Schenk

1. Thallus mostly many celled, holocarpic, endobiotic on *Spirogyra,* pollen grains, mosquito larvae and crab eggs.

Leptomitus Agardh

1. Oogonia absent, sporangia hypha-like, terminal.
2. Thallus with basal hold fast bearing with characteristics constriction of aseptate erect hyphae. Cell wall consists of chitin.

Pythium Pringsheim

1. Saprophytic or facultative parasite, which mainly cause root rot or damping off disease of seedlings.
2. Mycelium, branched, hyaline, coenocytic,multinucleate and vacuolated.
3. Sporangia, globose or oval, arise apically.
4. Zoospores reniform, laterally biflagellated.
5. *P. aphanidermatum* (root rot of papaya), *P. debaryanum, P. ultimum, P. irregulare* (all cause damping off disease).

Phytophthora de Bary

1. Soil inhabitant, saprophytic or facultative parasite.
2. Mycelium branched and coenocytic, sporangia on specialized hyphal sporangiophores which are branched and with indeterminate growth.
3. Zoospores reniform, laterally biflagelated.
4. *P. infestans* (late blight of tomato and potato), *P. colocasiae* (leaf blight and corm rot of colocasia).

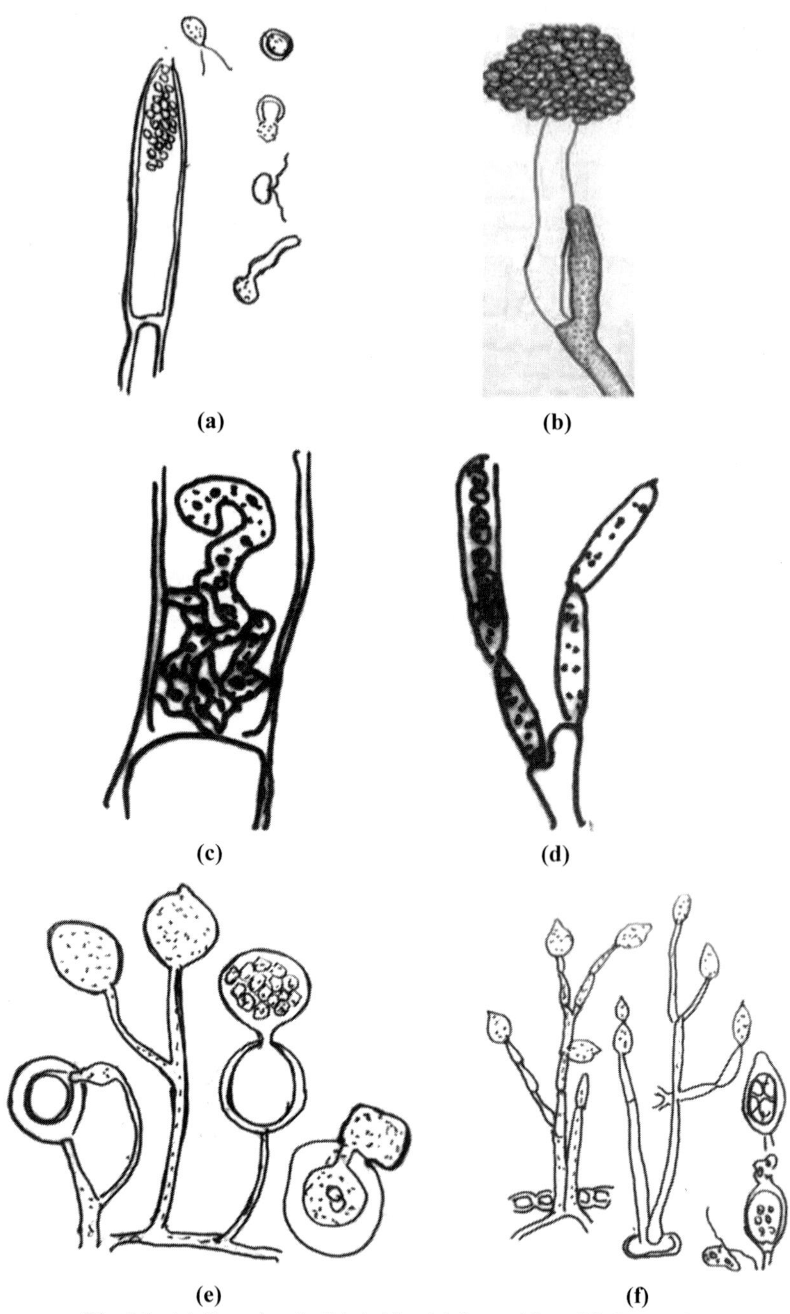

Fig.6.6. **(a)** *Saprolegnia* **(b)** *Achlya* **(c)** *Lagenidium* **(d)** *Leptomitus* **(e)** *Pythium* **(f)** *Phytophthora*

Sclerophthora* Thiruem *et. al.

1. The sporangiophore with citriform sporangia are like *Phytophthora*, while thick walled oospore is similar to *Sclerospora*.
2. It is an obligate parasite, cause downy mildew on cereals and grasses.
3. *S.rayssiae* var. *zeae* on maize and *S. cryophila* on orchard and forage grasses.

***Sclerospora* Schroet**

1. *S.graminicola* causes green ear disease of bajra.
2. Sporangiophores swollen with finger like projection ending with hyaline sporangia.

***Plasmopara* Schroet**

1. *P.viticola* causes downy mildew of grapes, obligate parasite.
2. Mycelium intercellular, haustoria simple or branched.
3. Sporangiophores in clusters on lower side of the infection spot.
4. Branches of sporangiophores arise at right angles and irregularly spaced. Final branches are known as sterigmata which ends with single sporangium.

***Peronospora* Corda**

1. A downy mildew fungus, differs from *Plasmopara* in having dichotomously branched pointed sterigmata.
2. Mycelium intercellular, haustoria simple or branched.
3. Sporangiophore in clusters on lower side of the infection spot

***Basidiophora* Roze & Cornu**

1. Sporangiophores unbranched, club shaped (basidium like) with swollen heads on which papillate sporangia present.

***Bremia* Regal**

1. Sporangiophores dichotomously branched, bear disc like structures with four sporangia on sterigmata.

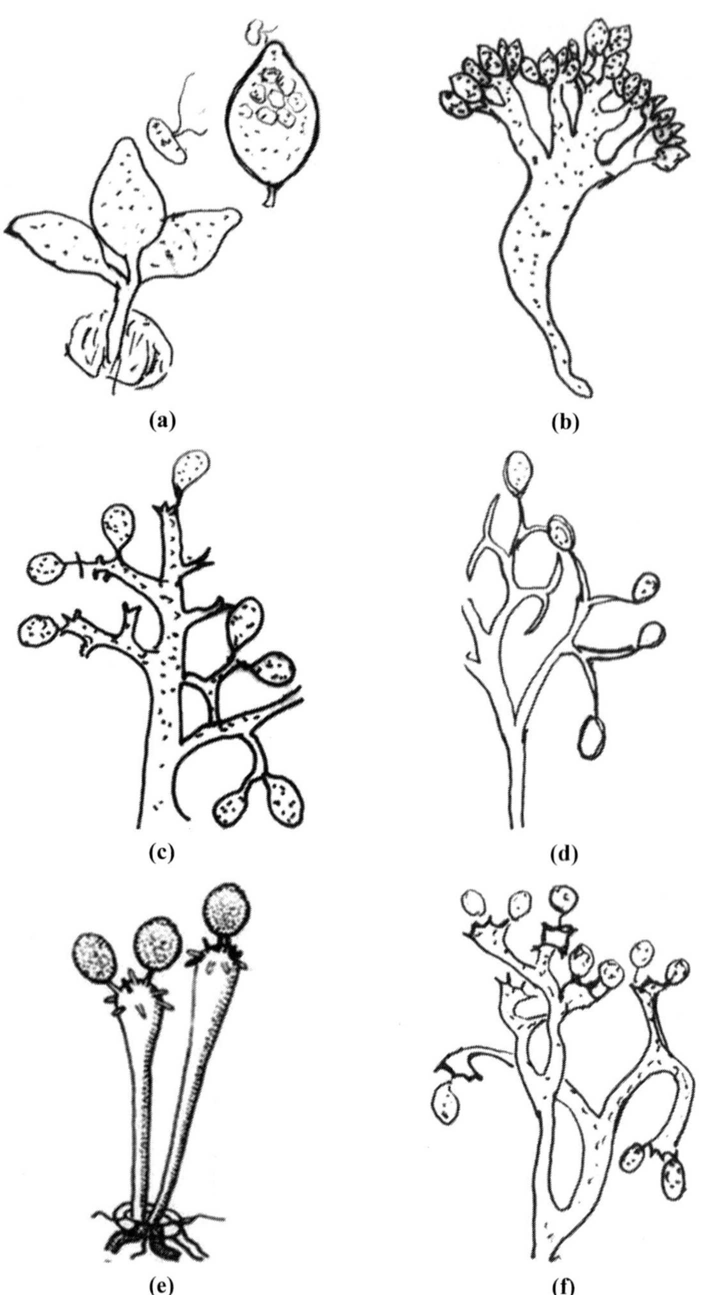

Fig. 6.7. Sporangiophore of **(a)** *Sclerophthora* **(b)** *Sclerospora* **(c)** *Plasmopara* **(d)** *Peronospora* **(e)** *Basidiophora* **(f)** *Bremia*

Albugo Pres. ex. Gray

1. About 30 species, obligate parasites, causes white rust disease.
2. Infected plant show white blisters, hypertrophy on different parts, mainly leaves.
3. Club shaped, stout sporangiophores are present in the infected area, below the epidermis.
4. Chain of sporangia is formed at the apex of sporangiophores.
5. On breaking off of epidermis, single celled, spherical, hyaline sporangia get detached from the sporangiophores which are the white rust stage.
6. Sexual reproduction takes place by forming antheridia and oogonia terminally on somatic hyphae.
7. On fertilization, oospores are formed. *A.candida* (on crucifers), *A.portulacae* (on *Portulaca oleracae*), *A. bliti* (on Amaranthaceae), *A. tragopogonis* (on *Tragopogon* sp.), *A. ipomoeae* (sweet potato, *Ipomoea*), *A. evolvuli* (*Evolvulus* sp.) *A. occidentalis* on spinach.

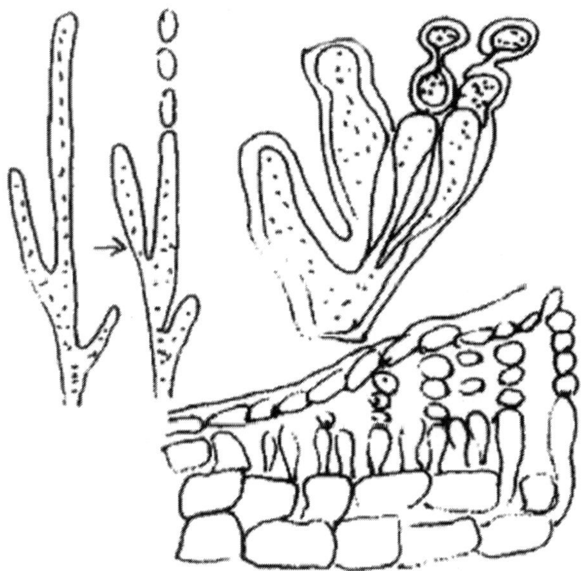

Fig. 6.8. Sporangia and sporangiophore of *Albugo*

6.3 Sub Division – Zygomycotina

6.3.1 Class – Zygomycetes

***Mucor* Micheli**

1. Present in soil, dung and other organic matter.
2. Rhizoids absent.
3. Sporangiophore arises singly from mycelium, mostly unbranched but rarely branched bearing sporangia on all branch ends.
4. Sporangia spherical, many spored, columella present.
5. *M.racemosus, M.globosus* are common.

***Absidia* Van Tieghem**

1. Sporangiophores straight, rarely single, more often in group of two to five, occurring on internodes of mycelium but not at the point of rhizoids (nodes).
2. Sporangia pyriform with columella.
3. *A. spinosa, A. heterospora, A. butler*i are common in soil.

***Rhizopus* Ehrenberg**

1. Cosmopolitan on soil, food or decaying organic matter. *R. stolonifer* (bread mold), *R. arrhizus*, *R.oryzae, R. cohnii* (all are lactic acid producers).
2. Mycelium of two kinds, one submerged in the substratum and the other aerial.
3. Sporangiophores arise opposite to the rhizoids and stolon.
4. Columella present.
5. Spores single celled, spherical, light brown in colour.

***Actinomucor* Schostak**

1. Nonthermophillic, rhizoids and stolon present.
2. *A. elegans* is used for making sufu or chinese cheese from soybean.

***Zygorhynchus* Vuill**

1. Soil inhabitant, homothallic Mucor-like columellate sporangia but differs from *Mucor* in its homothallic zygospores that are produced from unequal suspensors.

***Pilobolus Tode* (shot gun fungus)**

1. A coprophilous fungus grows on dung.
2. Mycelium coenocytic, branched in the substratum.
3. Sporangiophores erect arise from carotene rich trophocyst.

4. Sporangium at the tip of sporangiophore, black button shaped, columullate.
5. Spores, 15–30 thousand per sporangium, thrown violently upto a distance of 2–2.5 meters.

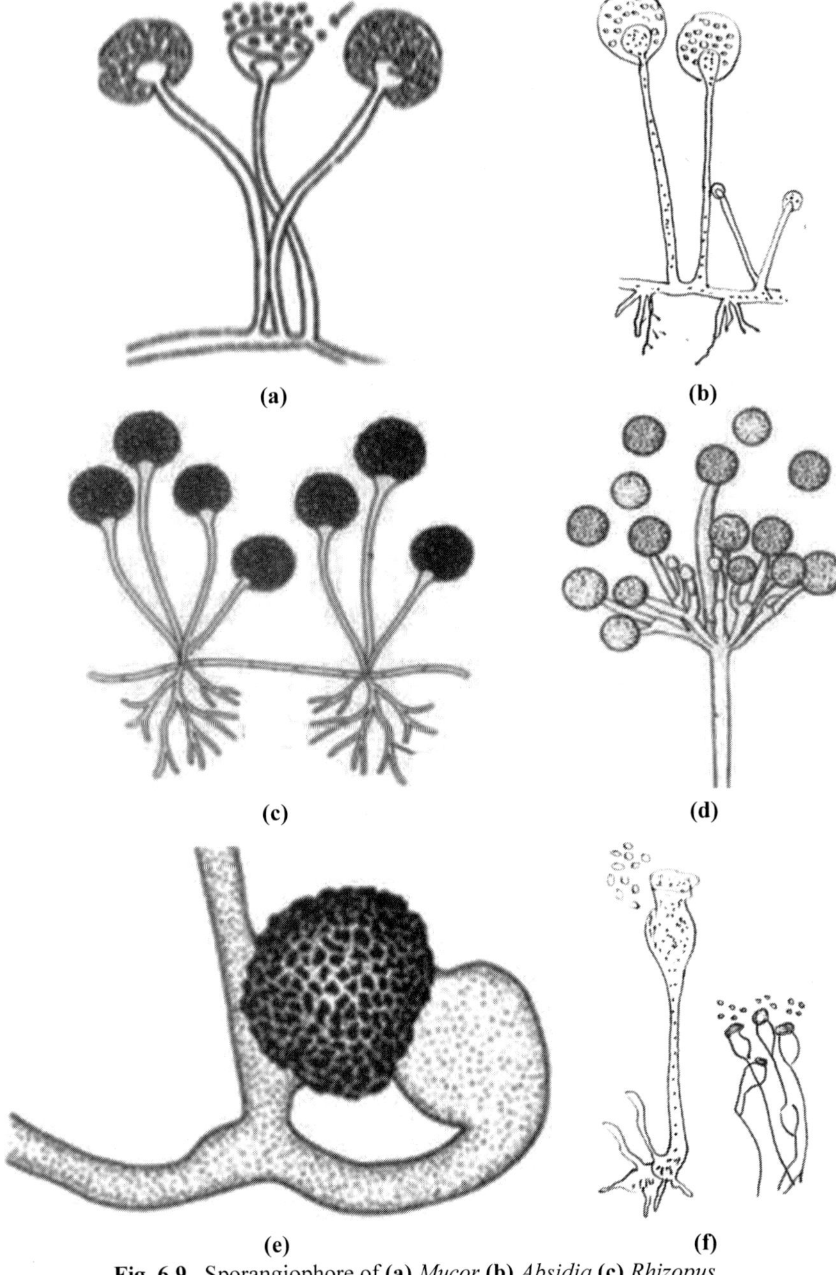

Fig. 6.9. Sporangiophore of **(a)** *Mucor* **(b)** *Absidia* **(c)** *Rhizopus* **(d)** *Actinomucor* **(e)** *Zygorhynchus* **(f)** *Pilobolus*

***Mortierella* Coemens**

1. Mycelium thin, delicate, branched.
2. Sporangiophores erect or branched.
3. Sporangia terminal, spherical, without columella.
4. Spores spherical or ellipsoidal.
5. Zygospores spherical, thick walled.
6. Common species *M.renispora, M.turficota.*

***Saksenaea* Saksena**

1. Flask shaped sporangia with spherical venter and long neck, columella prominent and dome shaped.
2. Spores single celled, spherical.
3. Sporangiophores, serial on dichotomously branched rhizoids.

***Helicocephalum* Corda**

1. Facultative parasite on nematode eggs.
2. Conidia large, brown to black (monosporus sporangiola) produced at the tips of unbranched conidiosphores with rhizoids at the base.

***Syncephalastrum* Schroeter**

1. Mycelium widespread in and on substratum without rhizoides.
2. Sporangiophores with rich sympodial branching.
3. Sporangia bearing heads globose, separated from sporangiophore by a wall.
4. Mero-sporangia (sporangioles) are finger like with many spores without basal cell.

***Piptocephalis* de Bary**

1. Obligate parasite on other members of Mucorales.
2. Sporangiophores dichotomously branched, cylindrical, merosporangiferous, vesicles small and deciduous.

***Syncephalis* Van. Tieghem & Le Monnier**

1. Mycelium inconspicuous in the substratum.
2. Conidiophores unbranched, swollen at the tip with mero sporangia, attached to the substratum by rhizoid like hyphae.
3. Spores cylindrical to cask- shaped, 1-celled.

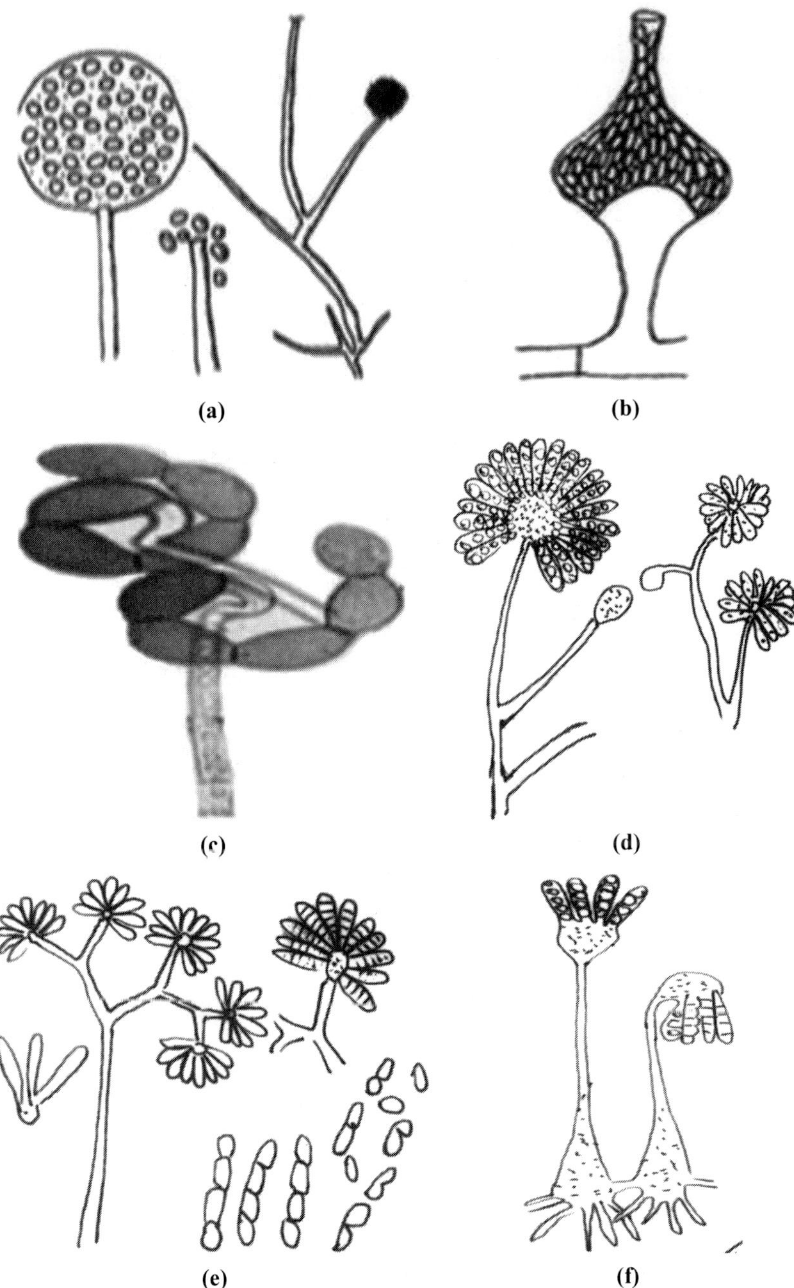

Fig. 6.10. Sporangia of **(a)** *Mortierella* **(b)** *Saksenaea* **(c)** *Helicocephalum* **(d)** *Syncephalastrum* **(e)** *Piptocephalis* **(f)** *Syncephalis*

***Dimargaris* Van Tieghem**

1. Obligate parasites on other mucorales, known to produce linear two-spored merosporangia on branched sporangiophores.
2. All sporangiophores are fertile.

***Kickxella* Coemans**

1. Sporangiophores erect, sporocladia sessile on an apical disc.
2. Sporangiola monosporous.

***Coemansia* Van Tieghem & Le Monnier**

1. Mycelium turf yellow, upto 6mm high, creeping, branched, septate.
2. Conidiophores unbranched, or forked with many septa.
3. Mero-sporangia alternating boat- shaped, broad, many celled, carrying on their inner surface a number of small basal cells which bear spindle shaped conidia.

***Cokeromyces* Shanor**

1. Terminal sporangia absent, lateral sporagiolae, spherical without columella.
2. On lateral stalks arising on swollen part of recurved sporangiophores.

***Entomophthora* Fr**

1. The insect destructor. About 82 species parasitic on arthropods and insects. *E.muscae* a house fly fungus.
2. The fly shows white bands of conidiophores on its abdomen.
3. Mycelium breaks into multinucleate hyphal bodies, each capable ofbecoming a sporogenous cell (conidiophore) which bears one or more asexual conidia.

***Basidiobolus* Eidans**

1. Saprobic in dung and gut contents of amphibians and reptiles.
2. Mycelium septate that does not break up into hyphal bodies.
3. Apex of the sporangiophore initially swells into holoblastic sporangium containing endogenous sporangiospore.
4. *B.ranarum* has large nuclei (upto 25 μm) good material for cytological studies.

***Zoopage* Drech**

1. Parasitic on protozoans.
2. Mycelium outside the host, non-septate, with lobed haustoria.
3. Conidia in chains on short projection.

Cochlonema Drech

1. About 18 species, endoparasitic on amoeba.
2. Thallus is coiled, branched or unbranched, forms chain of spindle – shaped conidia.

Fig. 6.11. **(a)** *Dimargaris* **(b)** *Kickxella* **(c)** *Coemansia* **(d)** *Cokeromyces* **(e)** *Entomophthora* **(f)** *Basidiobolus* **(g)** *Zoopage* **(h)** *Cochlonema*

6.3.2 Class – Trichomycetes

***Harpella* Leger & Duboscq**

1. Thallus unbranched, septate on immature insects.
2. Trichospores are characteristic of the group, have four appendages, cylindrical, slightly curved or coiled.

***Asellaria* Poisson**

1. Thallus attached to hindgut of insects, basal cell swollen, lobed, with rhizoids or tooth-like outgrowth.
2. Fragmentation of the thallus results into formation of arthrospores.

***Palavascia* Tuzet& Manier. Ex Lichtw**

1. Thallus coenocytic, produce one type of sporangiospores directly.

Amoebidium

1. Thallus aseptate, unbranched, on aquatic insects on outer surface of the host.
2. Spores either with rigid walls or amoeboid cells.

6.4 Sub Division (III): ASCOMYCOTINA

6.4.1. Class – Hemiascomycetes

***Protomyces* Unger**

1. *P.macrosporus* causes stem galls on coriander and other umbelliferae members.
2. Mycelium filamentous, septate, intracellular in the host.
3. The galls develop due to hypertrophy.
4. Swelling of hyphal cells results into formation of resting bodies.
5. These later on develop into sac like structures with numerous endospores (ascospores).
6. Two ascospores conjugate on the host to from diploid spores.
7. Only these spores are capable to infect the host.

***Ascoidea* Brefled**

1. About five species, grow on exudate of plants and ambrosia beetles.
2. Asci multispored.
3. Ascospores hat shaped and blastic conidia.

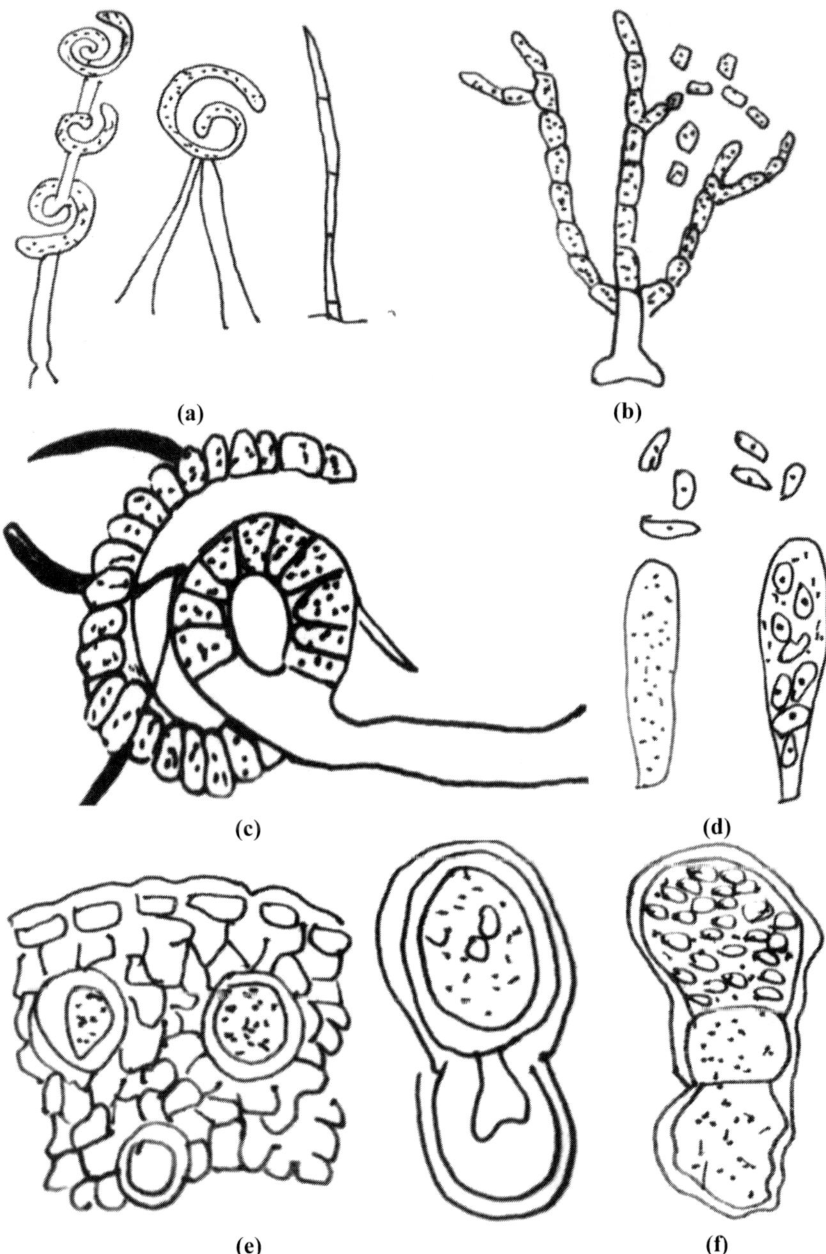

Fig. 6.12. Thallus of **(a)** *Harpella* **(b)** *Asellaria* **(c)** *Palavascia* **(d)** *Amoebidium* **(e)** *Protomyces* **(f)** *Ascoidea*

***Saccharomyces* Meyen (Gr.saccharon – sugar –myces = fungus)**

1. All the species have round ascospores and cause vigorous fermentation of sugars.
2. *S.cerevisiae* is the yeast of bakers and brewers.
3. It is used in the production of alcohol and in medicine for vitamin B-complex.
4. The thallus single celled multiply asexually by budding or by fission.
5. Life cycle haplo-diplobiontic type in which both phases are equally extensive and important.

***Spermophthora* Asbby& Nowell**

1. *S.gossypii* causes stigmatomycosis of cotton bolls and tomato fruits.
2. Thallus coenocytic filamentous.
3. Apical cell of the diploid hyphae develop into ascus with 8 spindles shaped ascospores.

***Dipodascus* Legarth**

1. Well developed septate hyhae.
2. Zygote mother cell develops into elongated ascus with numerous rounds to oval ascospores.

***Eremascus* Eidam**

1. Two species, *E. albus* and *E. fertilis* growing in sugary substrates.
2. Mycelium well developed, septate.
3. Zygote develops into round ascus with 8 ascospore.

***Taphrina* Fr.**

1. The species of *Taphrina* cause necrotic spots, galls, deformation and witches broom on plant.
2. Mycelium subepidermal and naked asci (no fruiting) present as parasitic phase in the host.
3. Ascospores soon after release, and also while still inside the asci, form blastospores by way of budding.
4. *T.maculans* on Turmeric, and *T.deformans* on Peach.

***Onygena* Pers. ex. Fr.**

1. Cleistothecia large sessile or stalked with pseudoparenchymatous peridium.
2. Grow on keratin rich substances.
3. Ascoma is stalked, cleistothecium globose.

4. Asci globose, scattered, dehisce and releases mass of dry ascospores.
5. Peridium is pscudopatenchymatous, made up of polygonal cells.

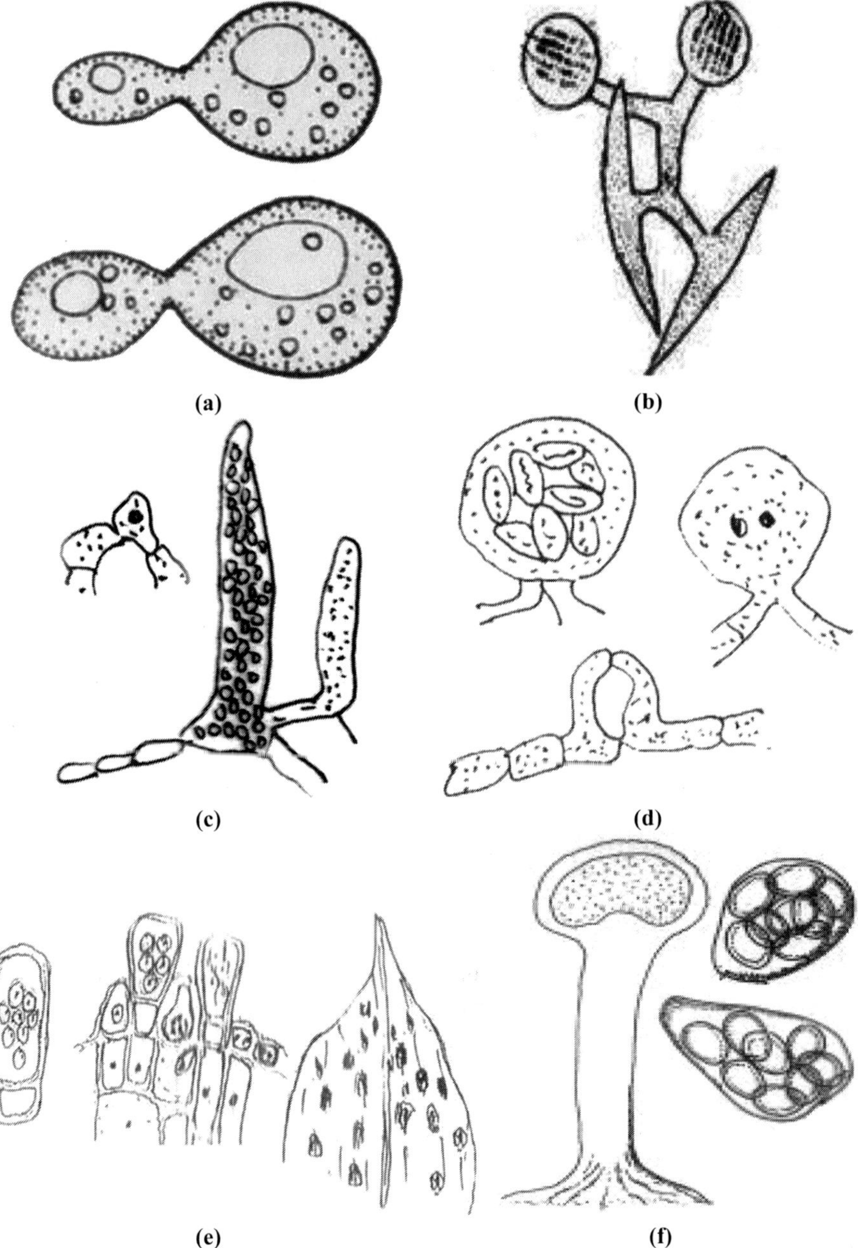

Fig. 6.13. **(a)** *Saccharomyces* **(b)** *Spermophthora* **(c)** *Dipodascus* **(d)** *Eremascus* **(e)** *Taphrina* **(f)** *Onygena*

***Eurotium* Link ex. Fr.**

1. The cleistothecium minute, globose, bright yellow, superficially on the mycelium.
2. The peridium is pseudoparenchymatous, single layer of polygonal cells.
3. The conidial stage is *Aspergillus glomus.*

***Emericella* Berk. & Br.**

1. The cleistothecium globose, reddish to purple, with several layers of cells, presence of thick walled hulle cells.
2. The conidial stage is *Aspergillus nidulans.*
3. Ascospores are pully wheel like, one celled.

***Sartorya* Vaill. & Benj.**

1. The cleistothecium white to pale yellow, surrounded by loose mass of sterile hyphae.
2. The conidial stage is *Aspergillus fumigatus.*

***Thielavia* Zopt.**

1. Cleistothecia globose, walls brown, pseudoparenchymatic, no appendages.
2. Asci oval, eight spored. Ascospores irregularly arranged, brown, one-celled.

***Ascosphaera* Alive & Spiltoir**

1. Ascomata larger.
2. Ascospores in balls (many spored asci), ascospores not spherical.

***Talaromyces* Benj**

1. Cleistothecia small, globose to sub-globose, non ostiolated, mostly yellow.
2. Asci numerous, ovate to globose, eranoscent. Ascospores mostly spiny, rarely smooth or girdled, hyaline (conidial stage, *Penicillium.*)

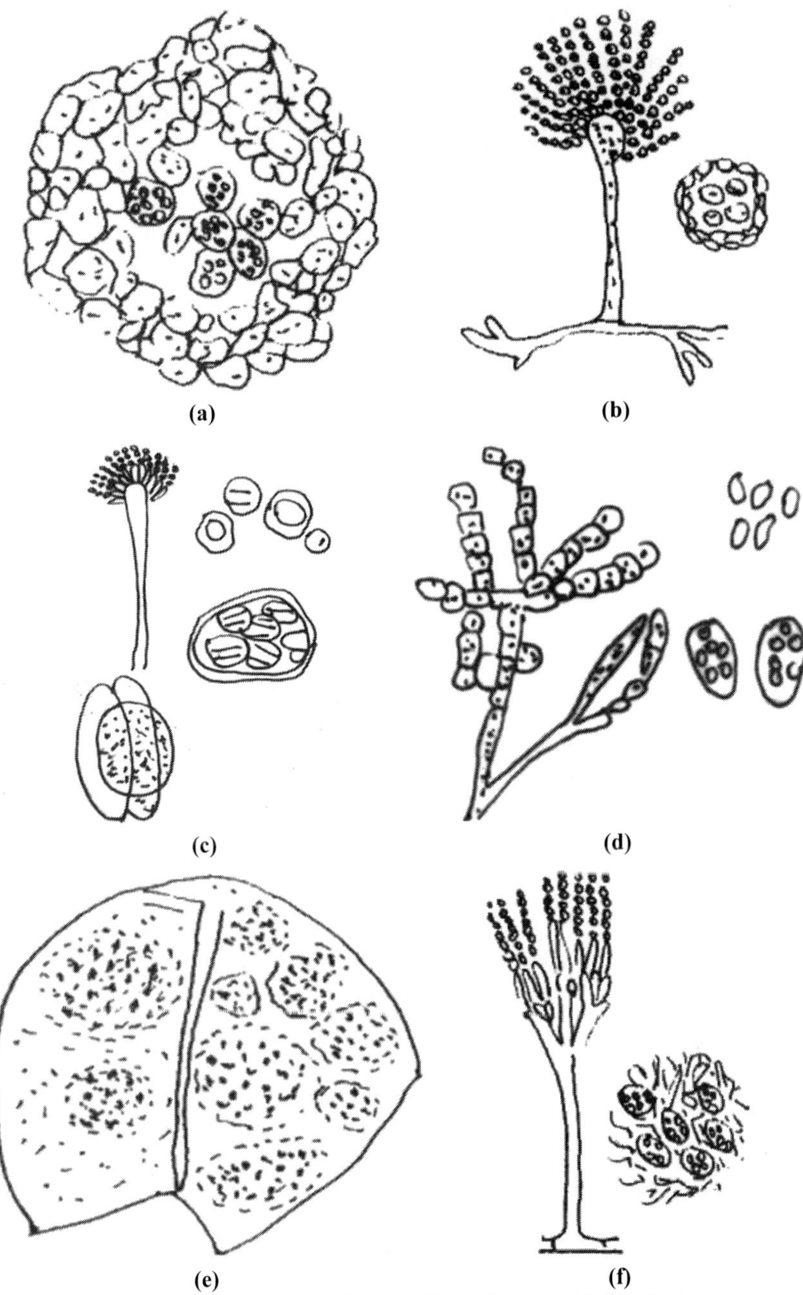

Fig. 6.14. **(a)** *Eurotium* **(b)** *Emericella* **(c)** *Sartorya* **(d)** *Thielavia* **(e)** *Ascosphaera* **(f)** *Talaromyces*

6.4.2 Class – Pyrenomycetes

Erysiphe Hedw. ex. Fr.

1. *Erysiphe* cause powdery mildew diseases on different plants.
2. Mycelium superficial, hyaline, septate and branched penetrating host tissue with haustoria.
3. Conidiophores, erect, unbranched, hyaline, septate, with chain of conidia.
4. Conidia enormous in number on leaf surface appear like a mass of white powder, hence the name "powdery mildew."
5. Conidia one celled, ellipsoidal or barrel-shaped.
6. Ascocarps (cleistothecia) superficial, brown to black in colour with myceloid simple and hyaline appendages.
7. Asci more than one.
8. Ascospores one- celled, hyaline. *E. graminis, tritici* (wheat) *E.graminis hordei* (barley) *E.polygoni* (several legumes).

Uncinula Lev.

1. *U.necator* and *U.tectonae* cause powdery mildews respectively on grape and teak plants.
2. Mycelium ectotrophic, hyaline, branched, penetration of host tissue with haustoria.
3. Conidiophores, hyaline, septate, unbranched, bearing one- celled conidia at the tip.
4. Ascocarps (cleistothecia) superficial, brown to black, appendages circinoid (coiled at tip).
5. Asci more than one per fruiting body.
6. Ascospores, single celled, hyaline.

Microsphaera Lev.

1. Ascocarps with dichotomously branched appendages having more than one ascus in each ascoma.
2. Fifteen species known. *M.grosulariae* on groseberry, *M.alphitoides* on oaks.

Phyllactinia Lev.

1. *P.corylea* causes powdery mildew on *Dalbergia sisoo.*
2. Mycelium hyaline, branched, partly internal (endophytic) in the form of globular haustoria.
3. Conidiophores emerge out through stomata, unbranched, rarely branched, wider than mycelium.

4. Conidia, one conidium per conidophore, one-celled, oval, pointed, and hyaline.
5. Ascocarp brown to black with appendages two types, one with swollen at the base and the other secretary appendages with broad base and apical branched with swollen tips.
6. Asci more than one, ascospores generally two per ascus.
7. Ascospores hyaline, one-celled.

***Sphaerotheca* Lev.**

1. Ascocarps with myceloid appendages, with one ascus in each ascocarp.
2. Imperfect stage is oidium. *S. pannosa* roses and *S. fuliginea* on melon and pumpkin.

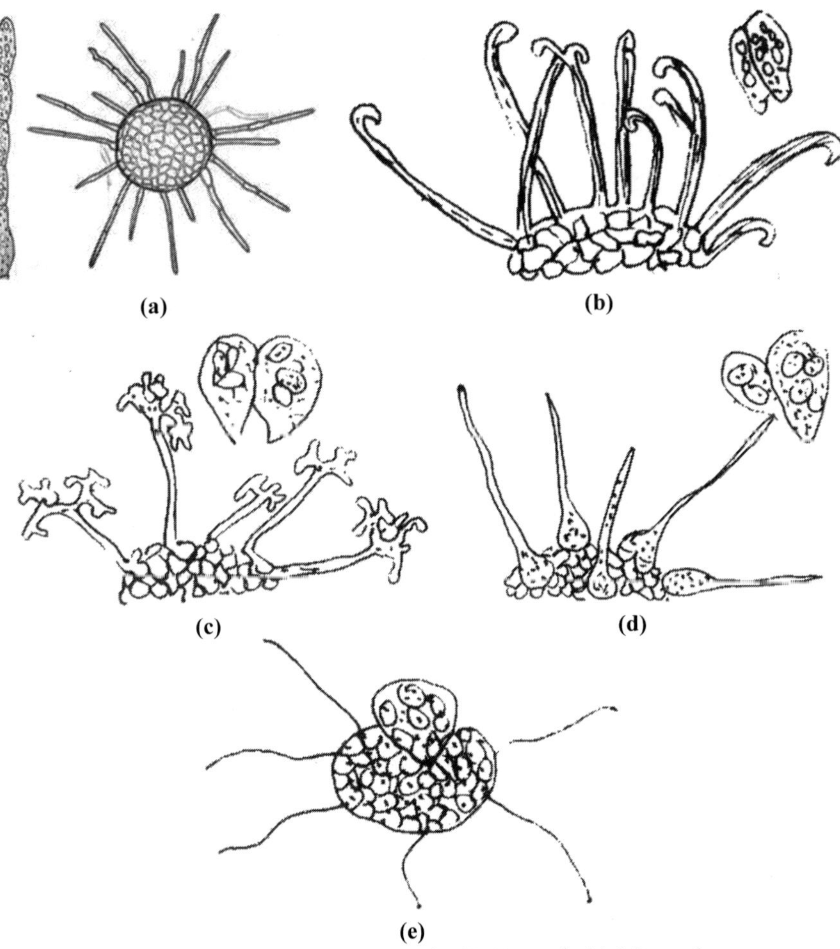

Fig. 6.15. Cleistothecia of **(a)** *Erysiphe* **(b)** *Uncinula* **(c)** *Microsphaera* **(d)** *Phyllactinia* **(e)** *Sphaerotheca*

***Podosphaera* Kunze. Ex. Luv**

1. Ascocarps with dichotomously branched appendages.
2. Single ascus in each ascocarp.
3. *P.leucotricha* (apple and other pome fruits), *P.oxyconthae* (plums, peach and apricots).

Meliola

1. Black or sooty mildew fungus found in tropical countries.
2. Characteristically black setae present on the mycelium.
3. Hyphae thick walled, brown, grow in patches on both the sides of infected leaves.
4. Hyphopodia originate from alternate cell of the hyphae in opposite directions.
5. Each hyphopodium consists of 2-cells, a short-basal cell and a swollen, club shaped or lobed head which serve as appresorium.
6. Ascocarp globose, black, attached to the mycelium.
7. Asci oblong, evanescent.
8. Ascospores 2-4 in each ascus, brown, elliptical, fusiform or slightly curved, 5-celled.
9. *M.jasminicola, M.argentina* and *M.palmicola* show germination on artificial media.

Coronophora

1. Ascocarps large, globose, black, smooth, immersed in the bark, singly or in clusters.
2. Asci clavate with long stalk.
3. Ascospores numerous, sausage-shaped, hyaline, unicellular.
4. *C.gregaria* grows saprophytically on the bark of deciduous trees.

Bertia De not.

1. *B.moriformis,* common on decorticated wood of *Pinus* and *Fergus,*
2. Forms dense cluster of ascocarps, which are black, resemble mulberry fruits.
3. Asci with 8 ascospores, Ascospores bicelled, sausage shaped, hyaline arranged in two irregular rows (bisireate).

***Ceratocystis* Eillis & Halst**

1. About 80 species known. *C.ulmi* (Dutchumetus disease) *C. fagacearun* (wilt of Oak), *C.fimbriata* (Black rot of sweet potato).
2. Asexual stages of the genus represent *Graphium, Cephalosporium*

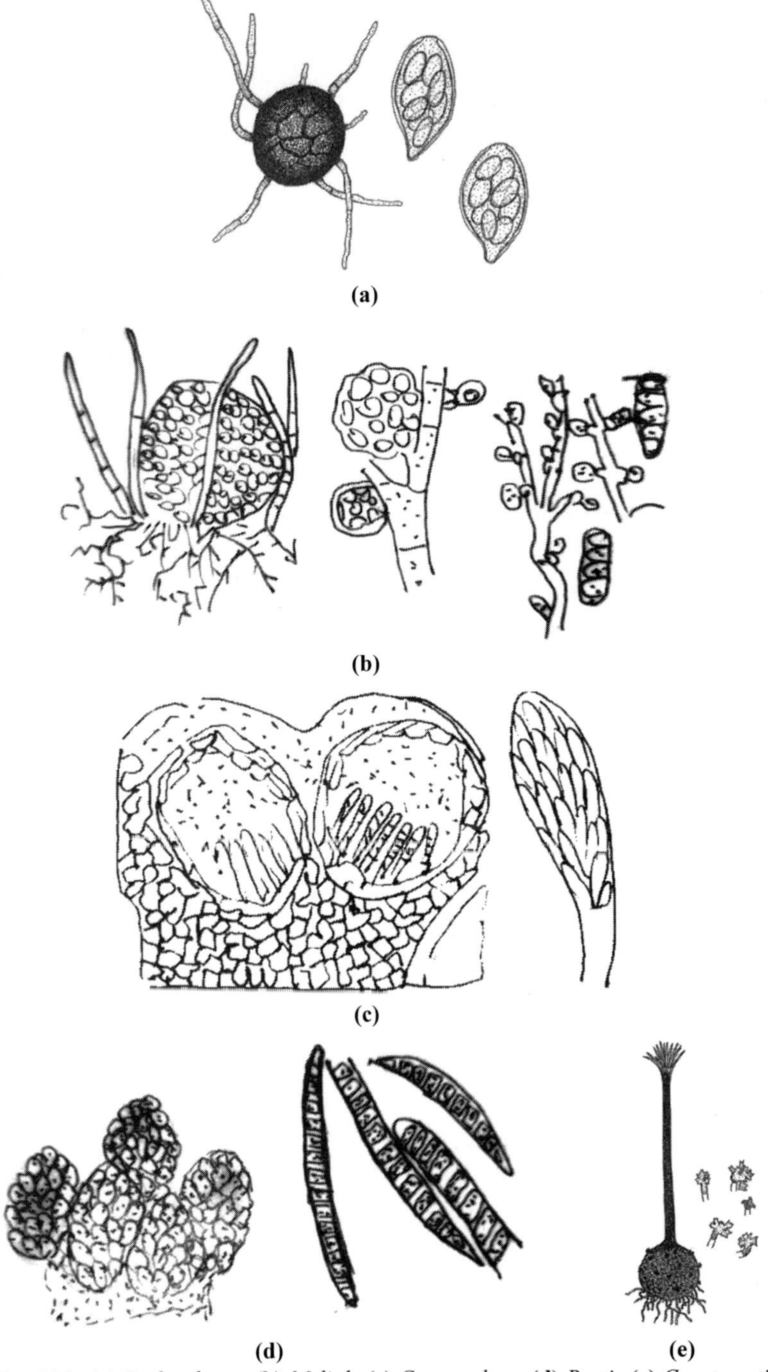

Fig. 6.16. **(a)** *Podosphaera* **(b)** *Meliola* **(c)** *Coronophora* **(d)** *Bertia* **(e)** *Ceratocystis*

Chaetomium Kunze and Schmidt

1. Mycelium superficial, thread like.
2. Perithecia, entirely superficial, on the mycelium, globose, ostriolate.
3. Peridium delicate, fragile forming head of characteristically branching hairs (mostly circinate).
4. Asci clavate, evanescent, 8-spored. Spores one-celled simple, hyaline to dark coloured.
5. Paraphyses present.
6. About 200 species, known for their highly celluloytic naure.
7. *C.thermophilum* is thermophillic and useful for compost preparation.
8. *C.globosum* common every where.

Hypocrea Fr.

1. Perithecia brightly coloured, fleshy, ostiolate, embedded in flat cushion-shaped sessile stromata, growing on dead wood, bark or old fructifications of fungi.
2. Ascospores colourless, on maturity sixteen ascospores in each ascus, divided into two.
3. Asexual conidia phialosporess of different types representing form genera of Deuteromycotina viz., *Trichoderma*, *Cephalosporium* and *Gliocladium*.

Nectria Fr.

1. Perithecia fleshy, beaked, embedded in a cushion-shaped stroma growing saprophytically on bark or fallen twigs.
2. Some species are wound parasites.
3. Asci elongate, ascospores two-celled with pseudoparaphyses.

Gibberella Sacc.

1. Stroma erumpent and pseudoparenchymatous.
2. Perithecia blue or violet, globose, ostiole papillate.
3. Asci clavate, cylindrical, 8 spored with evanescent paraphyses.
4. Ascospore hyaline or highly pigmented, 2-more transverse septa.
5. The imperfect stage belongs to *Fusarium*,
6. About 13 species, *G.fujikuroi* (Bakanae disease of rice), *G. avenacea* (damping-off to many hosts).

Phyllachora Nitsch. & Fuckel.

1. The species cause tar spot disease on leaves of different plants.

2. True perithecia with definite walls but embedded in endophyllous stroma on infected leaves, present on both sides.
3. The stroma black, resembles rust pustules, globose, ostiole distinct.
4. Asci with paraphyses.
5. Ascospores eight, one-celled, spherical.

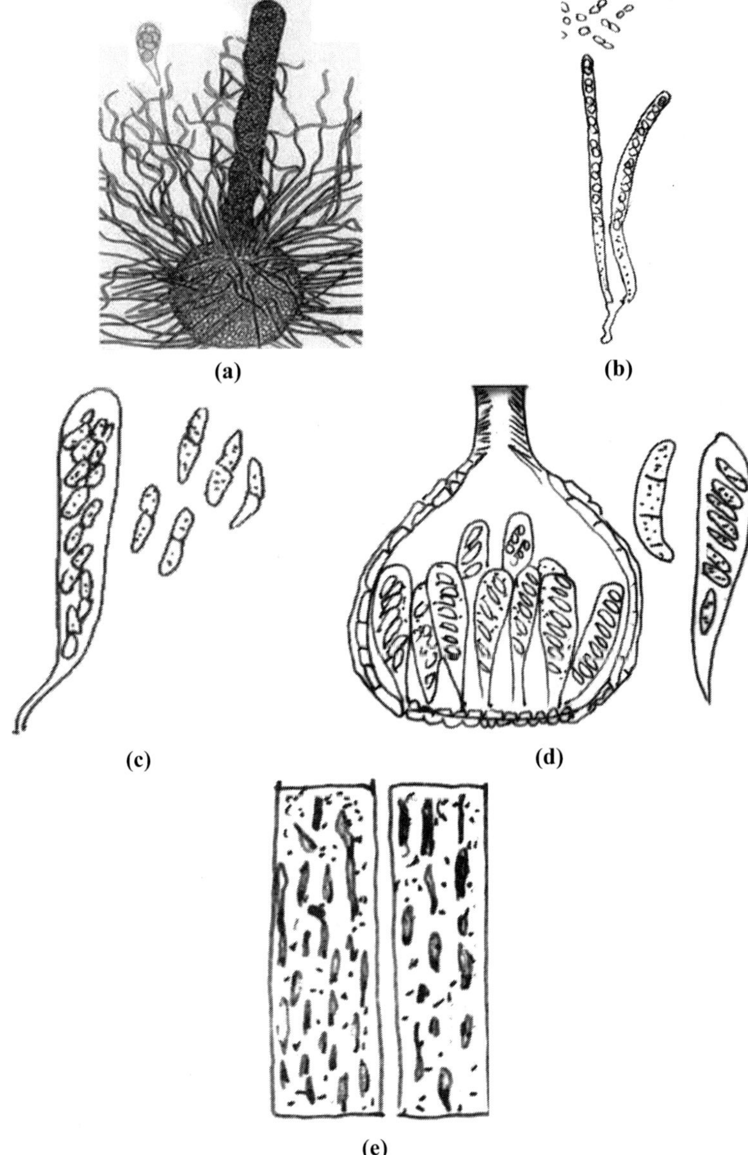

Fig. 6.17. **(a)** *Chaetomium* **(b)** *Hypocrea* **(c)** *Nectria* **(d)** *Gibberella* **(e)** *Phyllachora*

***Glomerella*: Spauld & Schrenk. C**

1. Perithecia beaked, in stroma or in host tissue.
2. Asci with short stalk, intermingled with paraphyses, 8-spored.
3. Ascospores one-celled, curved, hyaline (*Colletotrichum* and *Gloeosporium* species have been found imperfect stages).

***Sordaria Ces.* & de Not.**

1. Coprophilous, growing on dung of herbivore animals.
2. Perithecia dark, pear-shaped, 9.5 mm high, sunken in the substratum.
3. The asci in tufts, arise from the base.
4. Paraphyses deliquesce early, absent at maturity.
5. Ostiole lined with paraphyses.
6. At a time one ascus is ahead in growth from the rest, it elongates, and pushed through ostiole, ejects its ascospores violently into air and thus all empty asci shrinke and remain in the perithecium only.
7. Ascospores with minute germ pore at lower end, and surrounded by a gelatinuous sheath, so that spores stick to grass leaves and find their way into the alimentary canal of herbivore animals, where the spores receive stimulation for their germination.

***Podospora Ces* Hedwigia**

1. Coprophilous, growing on herbivore dung.
2. Perithecia pear-shaped, transparent, sunken in dung.
3. Ascospores 4-6 (variable) with mucilaginous-appendages on both ends.

***Neurospora Shear* & Dodge**

1. The name *Neurospora* due to presence of striations on the ascospores.
2. It grows on bread and burnt grounds. *N. sitophila*.
3. It is also known as "weed of the laboratory" for its fast contaminating nature.
4. The dry conidia in mass are pink in colour.
5. The mycelium highly branched, pigmented, Conidia two types, macro- conidia in branched chains, 6/8 μ diameter, multinucleate (*Monilia* stage) and micro-conidia 3-4 μ diameter, uninucleate, produced also in branched chains.
6. The mature perithecia, dark pyriform, beaked, no paraphyses,
7. Asci with 8-ascospores, which are brown to black with striations. *N.tetrasperma* only 4-ascospores per ascus.

***Xylaria* Hill**

1. Stroma stipitate, upright, corky leathery, fleshy, or woody.

2. Dirty white, brown or black externally but white internally.
3. Occur on dead rooting stumps of deciduous trees.
4. Stroma erect, club shaped, cylindrical, blackish below and whitish above.
5. Asexual (conidial) stage first occurs on upper surface of the stroma while after that perithecial stage occurs on lower portion of the same stroma.
6. Perithecia lie right angles to the surface, open externally.
7. Ascospores single celled.
8. *X. hypoxylon* (candle snuf fungus) and *X.polymorpha* (Dead mans's finger) are the two common species.

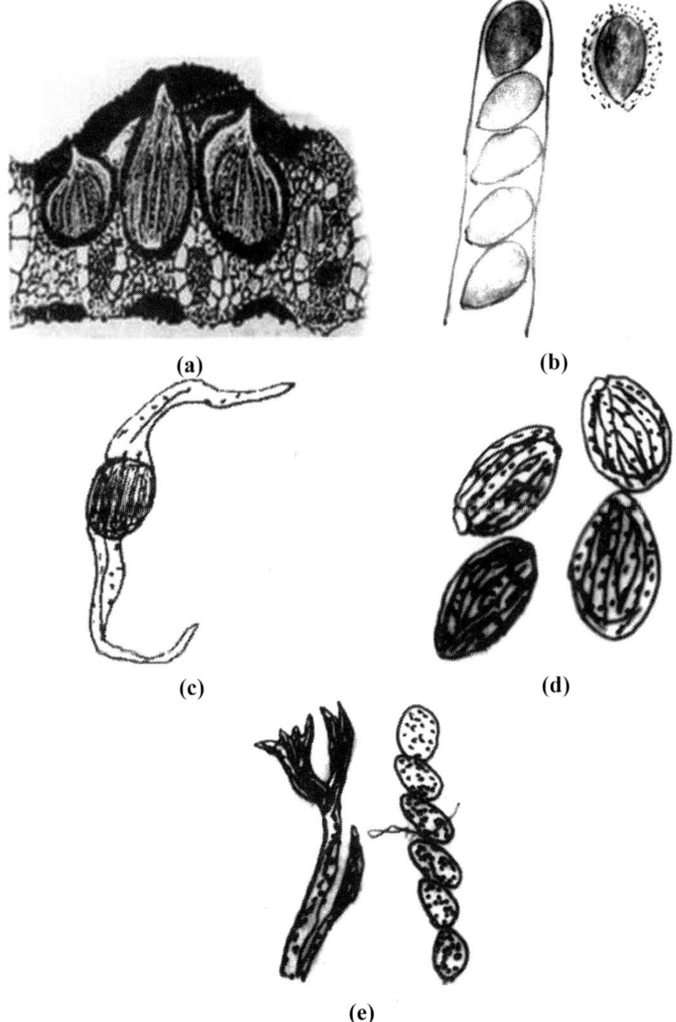

Fig. 6.18. (a) *Glomerella* **(b)** *Sordaria* **(c)** *Podospora* **(d)** *Neurospora* **(e)** *Xylaria*

***Claviceps* Tul**

1. It infects the inflorescence of grasses causing the disease ergot.
2. The two species *C.purpurea* (ergot of rye) and *C.microcephala* (ergot of bajra) are best known.
3. The sclerotium originates with the infection of host ovary, which is known as ergot body.
4. Erect stoma with long stalk, head globose arising from black sclerotium.
5. Perithecia immersed.
6. Honey dew stage on host consists single celled, hyaline conidial mass as asexual stage of the fungus.

***Cordyceps* (Fr.) Link**

1. An entomogenous, growing on insects and worms.
2. The fungus grows in and on the body of insects then the insect becomes sluggish and die.
3. The mycelium develops into sclerotium.
4. The dead insect enlarges in size and becomes resistant to decay due to antibiotics (cordycepin) produced by the fungus.
5. The dead insects with endosclerotium remains burried under ground. When favourable conditions for germination are available, club-shaped, orange stroma with a stalk and head arise above ground.
6. Perithecia in the periphery of stromatal head.
7. Asci with eight filiform ascospores, which multiply into small segments, disperse by wind and infect the insects.

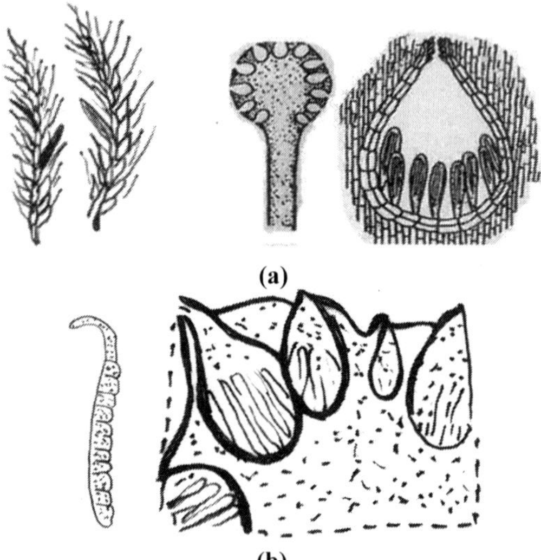

Fig. 6.19. (a) *Claviceps* **(b)** *Cordyceps*

6.4.3 Class – Discomycetes

***Medeolaria* Thaxt**

1. Parasitic on Indian cucumber root, resulting in shortening of internodes and fusiform swelling of the stem.
2. Ascocarps, indefinite.
3. Asci and paraphyses in the palisade layer.
4. Asci with 8- ascospores.

***Cyttaria* Berk.**

1. Parasitic on Fagaceae members, confined to south hemisphere.
2. Infected plants show galls on stem, bearing spherical stroma of the fungus and with numerous (about 200) embedded apothecia.
3. Asci with 8-ascospores.

***Tuber Mich.* Ex. Fr.**

1. Grows in mycorrhizal association (ectotrophic), ascocarps, with certain trees lie burried in soil, globose, ovoid, or irregular in shape, 2-3 cm in diameter with hard, black and warted rind.
2. Asci not organized in hymenium.
3. Ascospores reticulate or spiny, 2-3 per ascus.
4. The fruit bodies are highly prized as a delicious food.

***Peziza* L. ex. Amans.**

1. Apothecia sessile or feebly stalked, cup or disc- shaped, grow on dung or rotting wood in soil, Asci and paraphyses are photorophic and present in the cup surface towards light.
2. Asci with distinct amyloid ring at the apex which stains blue with iodine.
3. Ascospores large, ellipsoidal, discharge violently in one lot.

***Morchella* L. Amans.**

1. Sponge mushroom, highly edible fungus, popularly known as ‘Gucchi’.
2. Grows on humus rich soils.
3. Ascocarp consists a sponge like hollow pileus and a stalk.
4. The pileus has the network of broad, shallow depressions and ridges.
5. The depressions are lined with hymenium and ridges are sterile.

***Verpa* Swartz. Ex. Pers.**

1. Commonly known as bell morel due to its companulate pileus like a thimble over the apex of the hollow stipe.

***Rhytisma* Fr.**

1. In most of the species apothecia are in stroma.
2. Ascospores filiform to clavate.
3. *R. acerinum* causes Tar spot on mapple.
4. The hyphal spots appear on the leaves, which later turn black (stroma).
5. The stromatic hyphae secrete a black sticky substance which connect them and host tissues to form the black compact tar spot.

***Sclerotinia* Fuckel.**

1. Parasitic with narrow host range, causes crown rot of legumes, pink rot of celery, drop of lettuce, stalk rot of potatoes, timber rot of tomato, cottony rot of vegetables.
2. Cottony mycelium is converted into sclerotium as to perennate the fungus.
3. On favourable condition, the sclerotium germinates to form stalked cupulate apothecia.
4. The asexual (*Botrytis*) stage is present within the infected ovaries of flowers.

***Monilinia* Honey**

1. Cause fruit rot, leaf spot and die back diseases of plants belonging to Rosaceae and Ericaceae.
2. *M. fructicola*, *M.laxa* and *M.frutigena* cause brown rot of stone fruits like plums, almond, peaches, etc.
3. Asexual (conidial) stage on blossom or on fruits, forming Monilinia type branched conidiophores with chain of oval conidia.
4. Such infected fruits later dry up and shrive.
5. Hyphae alongwith host tissues form the stromatic stage and such mummified fruits lying burried in the soil for 1-3 years, give rise to long stalked, cupulate apothecia.
6. Ascospores hyaline. 1-celled, oval.

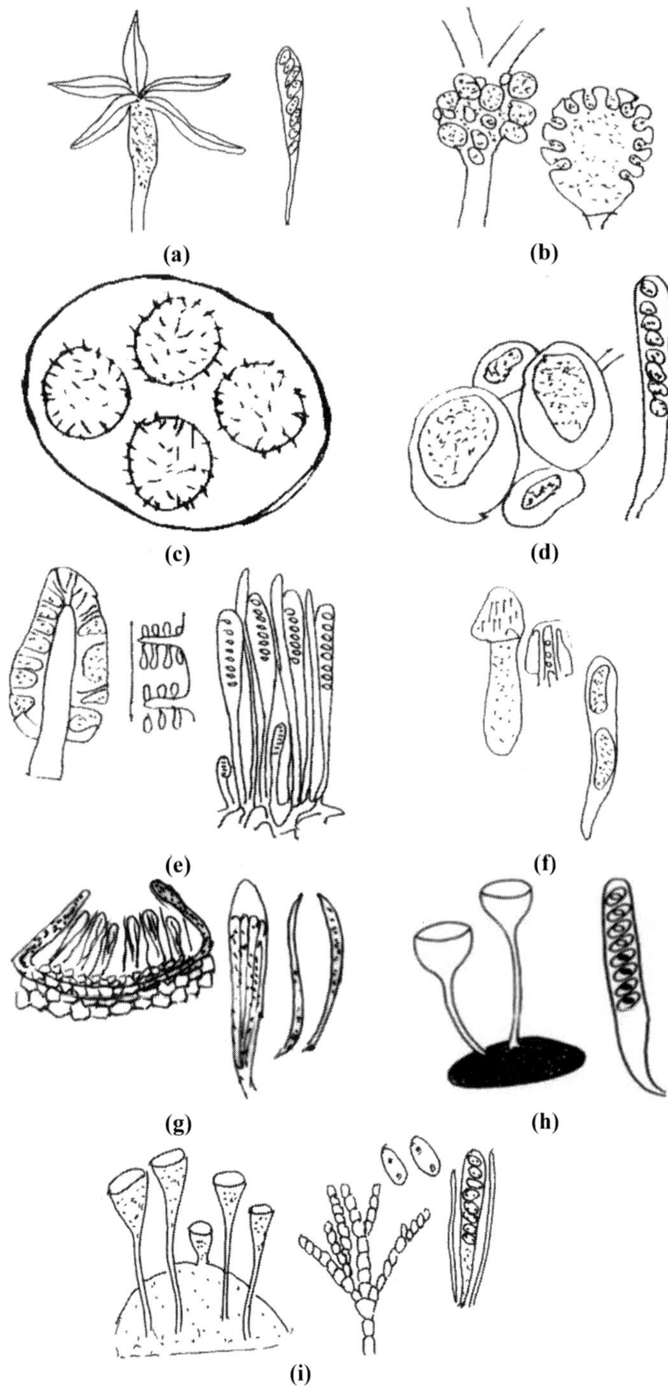

Fig. 6.20. **(a)** *Medeolaria* **(b)** *Cyttaria* **(c)** *Tuber* **(d)** *Peziza* **(e)** *Morchella* **(f)** *Verpa* **(g)** *Rhytisma* **(h)** *Sclerotinia* **(i)** *Monilinia*

6.4.4 Class- Laboulbeniomycetes

Zodiomyces Thaxter

1. Parasitic on species of Hydrophilidae (Coleoptera).
2. The receptacle is a multicellular, turbinate structure forming cup-like depression at the distal end which is bordered by numerous filamentous sterile appendages mixed with stalked perithecia and fertile branches bearing spermatia.
3. The spermatia are small, rod-like uninucleate and come in contact with trichogyne of ascosporangium.
4. The mature perithecium shows three basal cells and its wall is made with four vertical rows of cells arranged in several tiers.
5. Ascospores spindle shaped, bicelled, upper cell bigger than the lower cell.
6. With gelatinous sheath, ascospores helps to stick them on the body of insect (host).

6.4.5 Class- Loculoascomycetes

Myriangium Mont & Berk

1. Basal part of the stroma sterile, cushion-shaped portion bears the stalked, cupulate ascostroma, which are filled with globose asci.
2. Ascopores hyaline, dictyospore (transverse and longitudinal septation).

Elsinoe Racib.

1. Stroma simple with no differentiation into sterile and fertile portions, pulvinate or crust-like, formed beneath the host cuticle.
2. Ascospores four celled, hyaline surrounded by gelatinuous sheath.
3. Asexual stage (conidia, unicellular) in acervulus represents from genus *Sphaceloma,*
4. *E.veneta* causes anthracnose or gray blight of raspberries.

Mycosphaerella Johnson

1. Ascostroma, asci arranged in fascicle (fan shaped) at the base of stroma.
2. Ascospores eight, bi-celled.
3. *M.arachidicola* and *M.berkeleyi* are perfect stages of *Cercospora arachidicola* and *C.personata.*
4. *M.musicola* causes sigatoka diseases of banana.

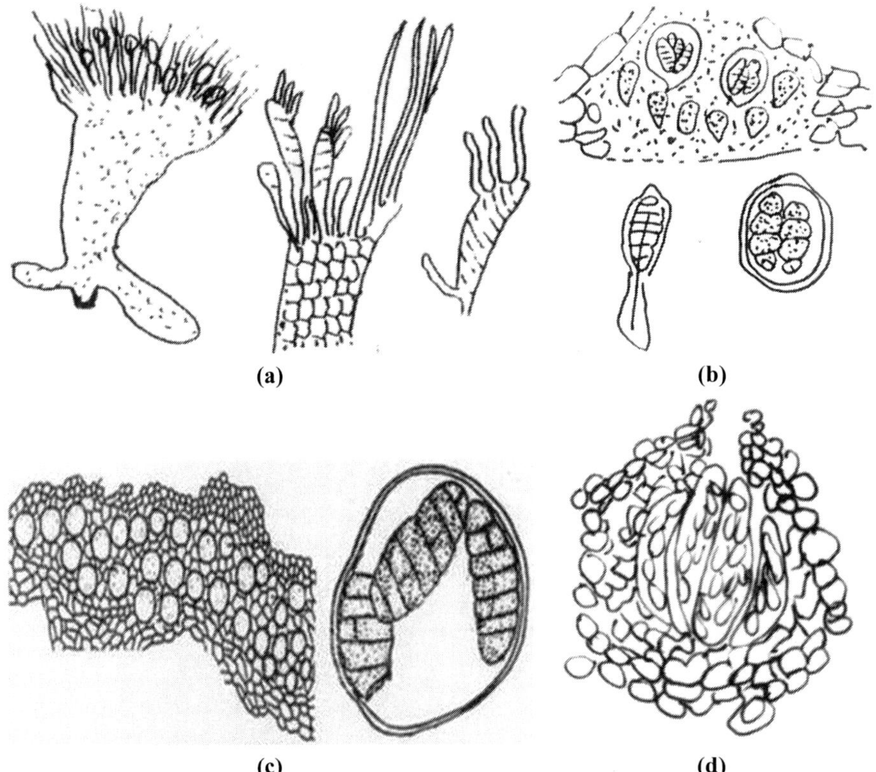

Fig. 6.21. **(a)** *Zodiomyces* **(b)** *Myriangium* **(c)** *Elsinoe* **(d)** *Mycosphaerella*

Pleospora Rabenh

1. Fruit bodies on dead herbaceous stems.
2. Well known for its muriform, pigmented, yellow to brown ascospores, in club shaped or bag shaped asci.
3. The conidial stage belonging to *Alternaria, Stemphylium or Phoma.*

Venturia Sacc.

1. Perithecoid ascostroma (Pleospora type centrum) remain immersed in the host tissue, open by an erumpent ostiole.
2. Asci arranged in between pseudoparaphyses which are attached to both the roof and floor of the locula.
3. Ascospores unequal, bi-celled.
4. The upper cell is broader than lower, imperfect stage is conidial to Hyphomycetes.
5. *V. inaequalis* cause apple scab disease.

Hysterium **Tode ex Fr.**

1. Grows on wood or bark forming black, elongated, longitudinally furrowed hysterothecia.
2. Ascospores brown, septate with distinct constrictions at the septa.

Microthyrium **Desm**

1. Ascocarp (thyriothecia) minute, superficial, circular with central pore on leaves.
2. Asci numerous, broad at base, tapering towards apex, 8-spored.
3. Ascospores irregularly arranged, fusiform, hyaline, bi-celled.

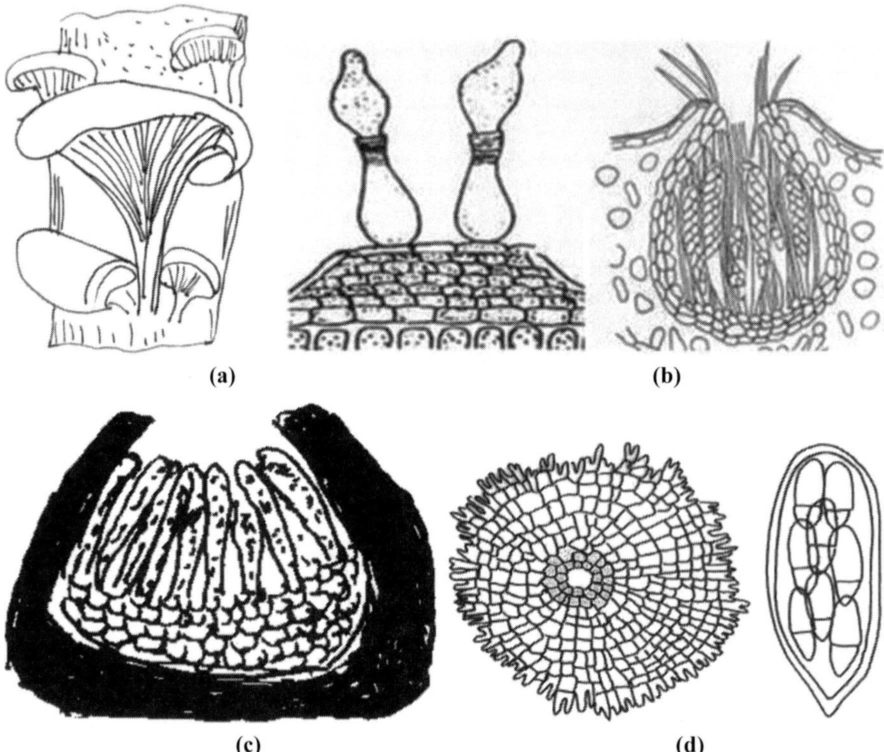

Fig. 6.22. **(a)** *Pleospora* **(b)** *Venturia* **(c)** *Hysterium* **(e)** *Microthyrium*

6.5 Sub Division Basidiomycotina

6.5.1 Class Teliomycetes

Puccinia **Pers.**

It is causing rust diseases on plants throughout the world, obligate parasite, autoecious or heteroecious, micro- or macro-cyclic.

1. Uredospore stage is known as rust stage of the diseases.
2. The uredinia are red in colour, appear below the host epidermis.
3. Uredospores oval, single celled with thick wall and stalked.
4. Teleutospores are formed in the black, elongated pustules.
5. Teleutospores spindle shaped, bicelled, dark brown and stalked.
6. Basidiospores are haploid, produced on germination of Teleutospores.
7. Pycniospores and aeciospores are produced on the same host in the autoecious rusts and on the alternate host in heteroećious rusts.
8. *P.graminis tritici* (black stem rust of wheat) *P.striiformis*-(stripe rust of wheat) *P.recondita*–(orange rust of wheat) *P.helianthi* (Sunflower rust) *P.penniseti*- (Bahra rust) *P.purpurea* – (Jowar rust). *P.arachidis* (Groundnut rust).

***Gymnosporangium* Hedwig ex. Dc**

1. Teliospores bicelled (rarely one or many celled) with pedicel which gelatinize on moist.
2. Restricted to temperate regions.
3. Most of the species are heteroecious, demicyclic (i.e., lack uredinia), the telial stage occurs on gymnosperm plants.
4. *G.juniperi virginianae* causes apple cedar rust.

***Uromyces* (Link) Unger.**

1. The second largest rust genus on members of Leguminosae.
2. *U.appandiculatus* on bean; *U.ciceris* on gram; *U.pisi* on pea.
3. One celled, stalked teliospores. (in *P.heterospora* has mixed teliospores single and bicelled).

***Hemileia* Bark and Br.**

1. Pycnia and aecia are not known.
2. Uredospores are formed in sub stomatal chambers.
3. The spores are reniform.
4. Teliospores 1-celled, thin walled, pale and germinate in situ without any resting period hence, termed leptospores.
5. *H.vastatrix* causes coffee rust.
6. The disease was reported in 1868 from Sri Lanka; It is an autoecious rust, pycnia and aecia lacking.
7. Uredospores and teliospores are formed in orange to brown coloured mixed sori in stomatal chambers at lower surface of leaf.

Phragmidium **Link.**

1. Teliospores stalked, transversely septate, multi-celled.
2. Autoecious macro- or demi-cycled on Rosaceae plants.
3. Teliospore has a long stalk surrounded by gelatinous sheath.

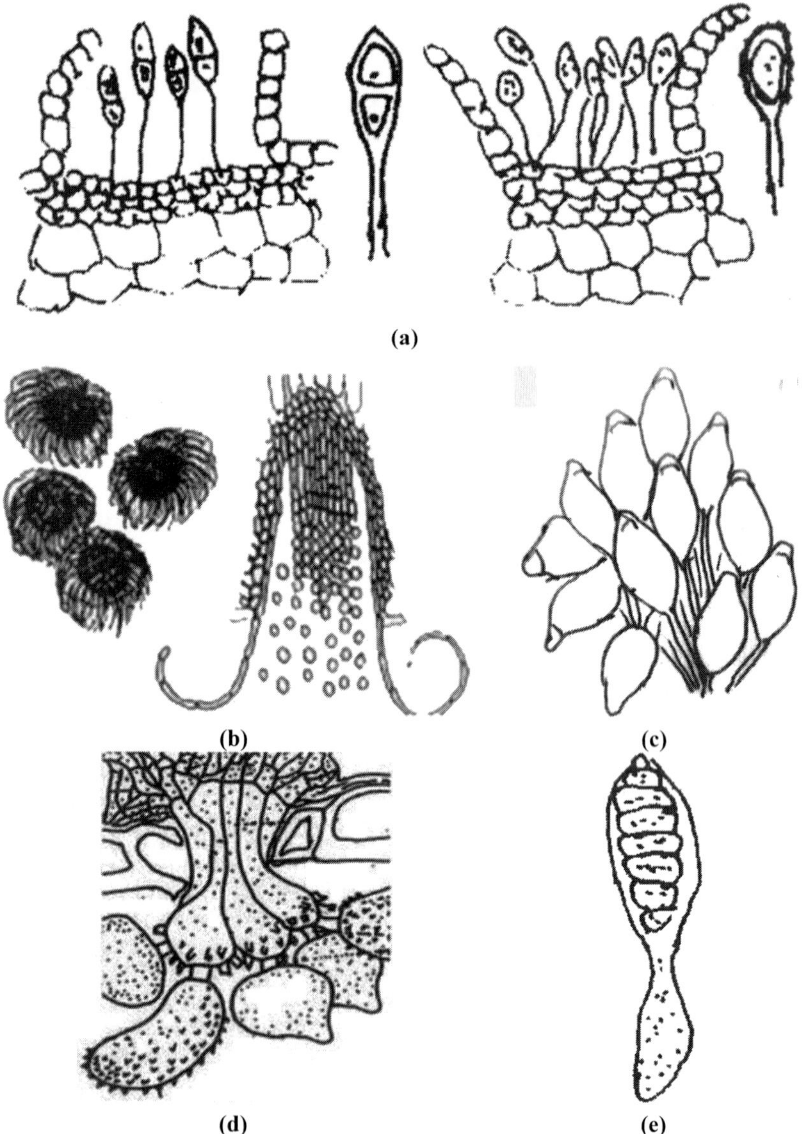

Fig. 6.23. **(a)** *Puccinia* **(b)** *Gymnosporangium* **(c)** *Uromyces* **(d)** *Hemileia* **(e)** *Phragmidium*

***Ravenelia* Berk**

1. Teliospores multicelled, compound head (formed by fusion of several teliospores, each with separate stalk).
2. Head smooth or ornamented with several small projection.
3. Uredospores one celled.
4. Rust of *Pongamea* sp.

***Ochrospora* Dietel**

1. Teliospores becoming 3 septated (internal basidia),
2. *O.ariae* on apple and pear.

***Melampsora* Cast**

1. Uredospores stalked, echinulate, brown with capitate paraphyses.
2. Teliospores single celled, sessile, formed under epidermis.
3. Pycnia flat and diffuse; aecia caeomoid
4. *M. lini* cause rust of linseed (flax)

***Masseeela* Dietel**

1. Presence of characteristic telial hair, projecting out of the infected leaves, which are gelatinous masses of detached 1- celled, sessile teliospores.
2. Macrocyclic and autoecious, on several members of Euphorbiaceae.

***Cronartium* Fr.**

1. Macrocyclic, heteroecious, spermogonia and aecia on Pinus.
2. Uredia and telia on Dicotyledons.
3. Teliospores sessile, fused to from finger-like columns on lower side of the leaf. *C.ribicola* causes (blister rust of Pinus).
4. The alternate hosts species are Ribes and goose berry.

***Melampsorella* Schroet.**

1. Heteroecious, producing aecia on *Abies* and *Picea.*
2. Causing witches brooms.
3. Teliospores laterally united, pale or colourless.

***Cerotelium* Arthur.**

1. Mostly in warm regions of the world.
2. *C fici* causes rust of fig.
3. Macrocyclic, teliospore in chain laterally.

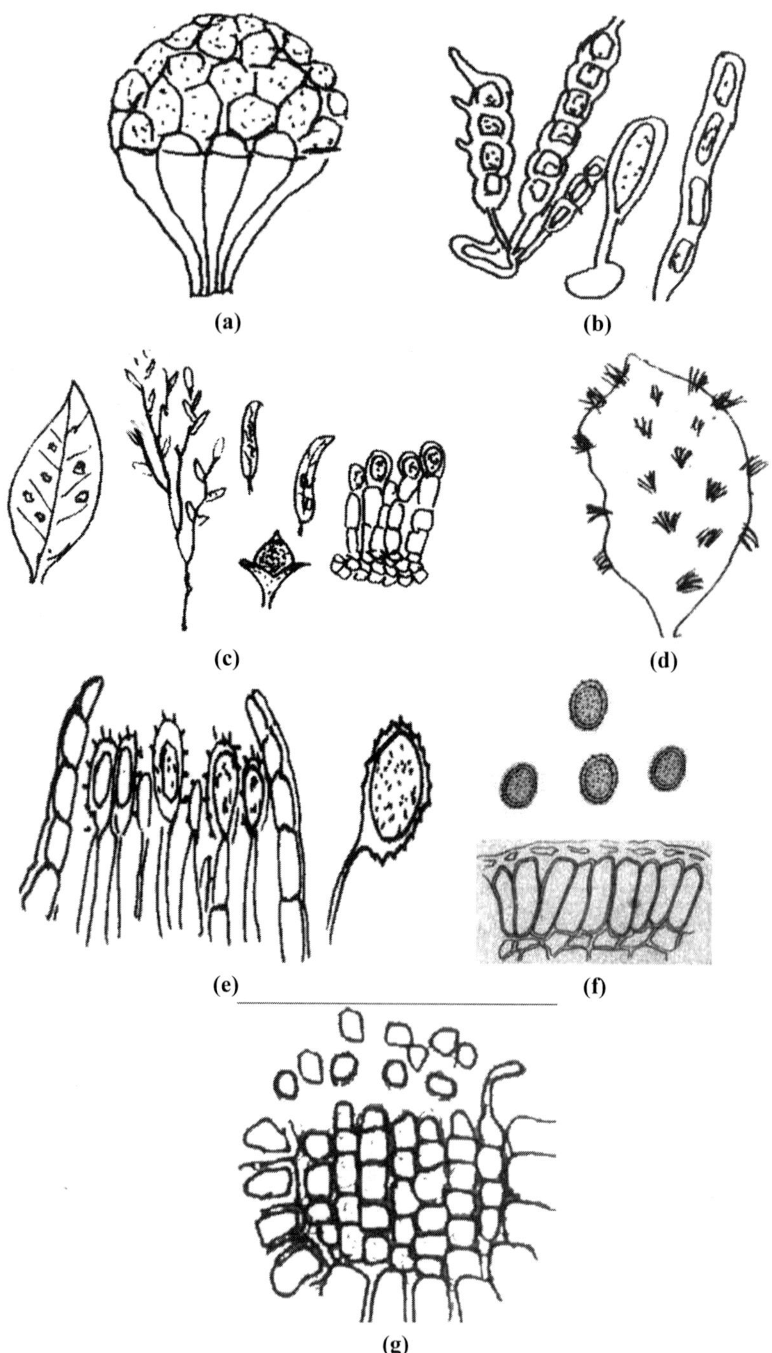

Fig. 6.24. **(a)** *Ravenelia* **(b)** *Ochrospora* **(c)** *Melampsora* **(d)** *Masseeela* **(e)** *Cronartium* **(f)** *Melampsorella* **(g)** *Cerotelium*

Dasturella

1. Telia black and hard with cells firmly fused together.
2. Causes rust on bamboo.

Smut Fungi

***Ustilago* Roussel**

1. Teliospores single celled, black dusty, peridum of host origin.
2. Columella absent.
3. Parasitic on members of Graminea *U.tritici* (loose smut of wheat).
4. *U.nuda* (loose smut of barley), *U. maydis* (maize smut), *U.scitaminea* (whip smut of sugarcane), *U.occidentalis* (smut of cyanodon).

***Sphacelotheca* de Barry.**

1. Teliospores single celled, in sori with peridium of fungal origin, and columella of host origin,
2. Spores black to dark brown, smooth, or ornamented, thick walled, mostly single.
3. *S.reiliana* (head smut of jowar), *S.sorghi* (grain smut of jowar). *S.cruenta* (long smut of jowar).

***Tolyposporium* Woron.**

1. Sori commonly in ovaries of host, spore balls covered by membrane of host origin.
2. Spores angular to round with roughened spore wall.
3. Germination of spores forming septate promycelium, bearing terminal and lateral basidiospores. *T.penicillariae* (smut of bajra).

Graphiola

1. Five species in tropical areas.
2. *G.phoenicis* parasitic on palm leaves.
3. Sori are formed under the epidermis and immediately emerge up tearing host tissues for exposing chains of spores (yellow).
4. Dark cup-like, hard periderm protects the central sporogenous cells.
5. The parallel chain of the spores (single celled) may bud laterally to form sporidia.
6. Sterile hyphae are also present.

***Tilletia* Tul**

1. Teliosori inside the infected grains, teliospores, brown, 1-celled, promycelium non-septate bearing 8-filiform sporidia at terminal end.
2. Exospore smooth, reticulate, verrucose, tuberculate or sculptured.
3. The sporidia form H-shaped structures.
4. *T.carries* and *T. foetida* cause bunt of wheat.

***Urocystis* Rabenh**

1. Sori on various parts of the host including roots, erumpent.
2. Spore balls with 1-4 dark fertile central spores surrounded by marginal sterile cells.
3. Flag smut of wheat (*U.tritici*) and onion smut (*U.cepulae*).

***Neovossia* Kornicke**

1. Teliospores, brown, or numerous, filifrom sporidia at the tip of promycelium.
2. Absence of H-shaped structure.
3. Karnal bunt of wheat, (*N. indica*) bunt of rice, (*N. horridia*)

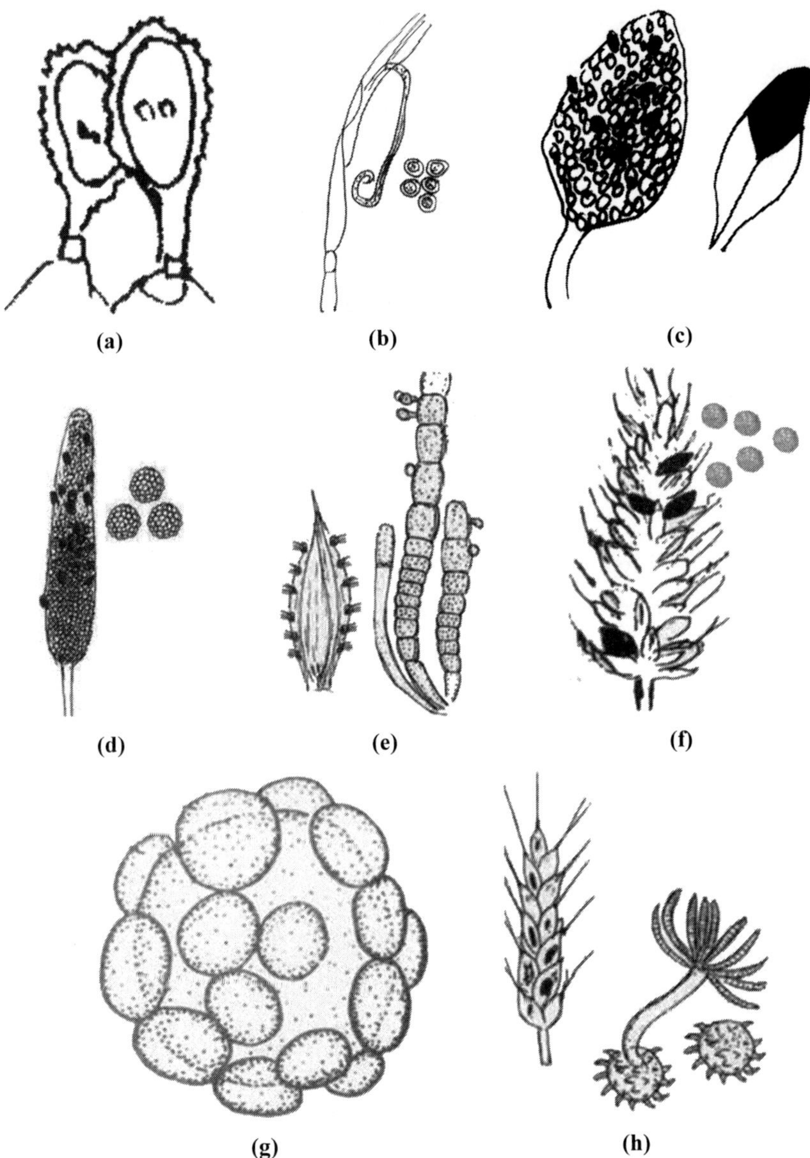

Fig. 6.25. **(a)** *Dasturella* **(b)** *Ustilago* **(c)** *Sphacelotheca* **(d)** *Tolyposporium* **(e)** *Graphiola* **(f)** *Tilletia* **(g)** *Urocystis* **(h)** *Neovossia*

6.5.2 Class – Hymenomycetes

Sub Class- Phragmobasidiomycetideae

Tremella Pers. Ex Amans.

1. Grow saprobically on tree trunks or decaying logs.
2. Basidiocarps cushion shaped or convoluted, gelatinous, variously coloured (white to bright yellow).
3. It is a large and cosmopolitan genus.
4. Basidiospores apiculate, borne asymmetrically on speculate sterigmata.
5. *T.reticulata* used as food in China.

Exidia Fr.

1. Basidiocarps variable in shape, tough gelatinous.
2. Basidia longitudinally septate.
3. *E.glandulosa* (witche's butter) black getatinous basidiocarps with small warts on decaying branches of woody hosts (Pine and Oak).
4. Hymenium on lower side of the basidiocarp.

Auricularia Bull ex. Merat

1. It is known as ear fungus (Auricular = ear), the basidiocarps resemble to a human ear.
2. *A.polytricha* edible, cultured and sold in the market of China.
3. Basidiocarp hairy on upper surface, gelationous center and broad hymenium on the lower side.
4. Basidia cylindrical, divide by three transverse septa.

Septobasidium Pat.

1. Symbiotic association with scale insects.
2. The fungus forms a mat of hyphae over the insect, which later develop into intricate basidiocarp, called a "fungal garden", the nutrition is drawn from insect.
3. The intricate basidiocarp is wonderful even for modern architects.
4. It consists of many vaulted chambers, arranged in several stories supported by pillers.

Sirobasidium

1. Basidiocarp gelatinous, folded or fan shaped.
2. Basidia catenulate, maturing basipetally, sterigate, sporoid and deciduous.

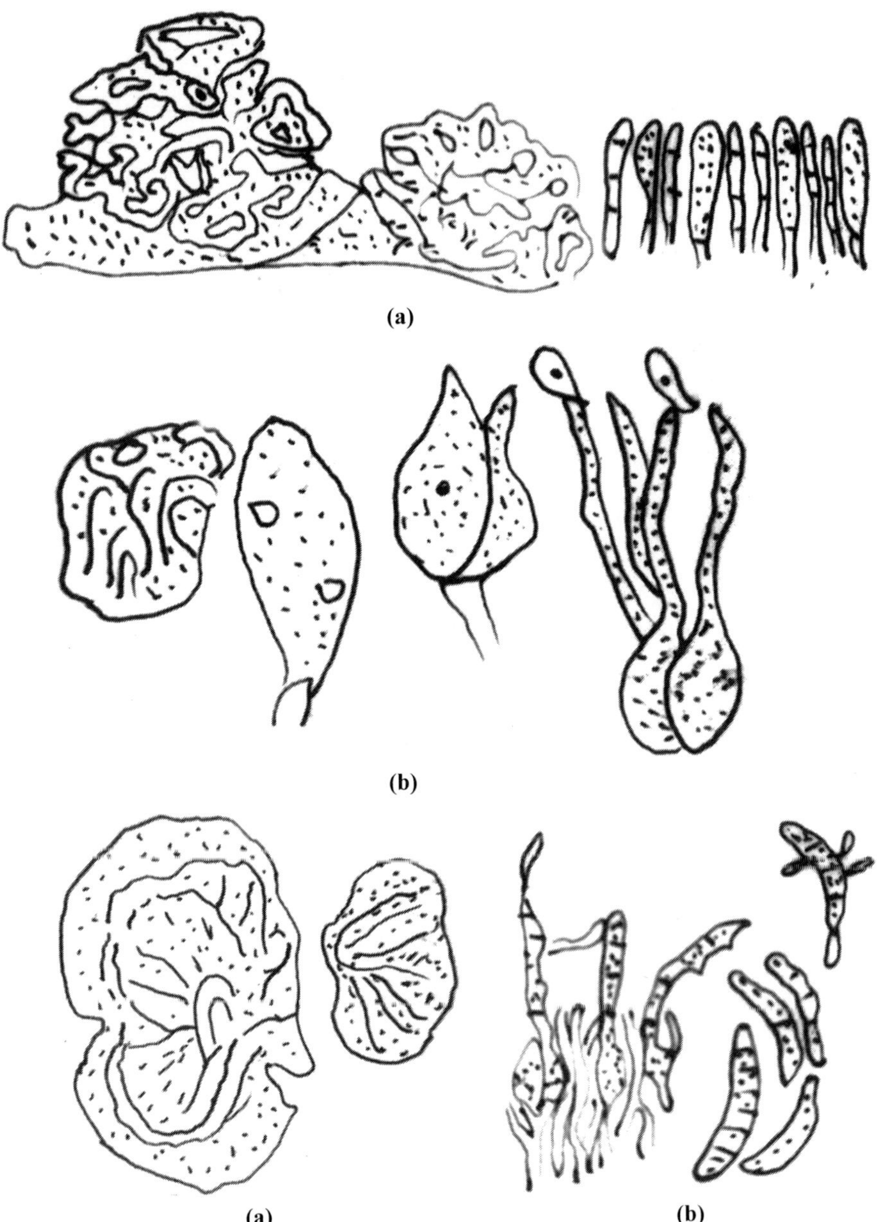

(a) (b)

Fig. 6.26. **(a)** *Tremella* **(b)** *Exidia* **(c)** *Auricularia* **(d)** *Septobasidium*

Sub Class- Holobasidiomycetidae

***Exobasidium* Woron.**

1. Parasitic on plants, absence of well defined basidiocarps (naked basidia) unspecialized holobasidium.
2. *E.vaccinii*, an obligate parasite on tea causing blister blight on leaves.
3. Causing hypertrophy and deformation.
4. Basidia are formed terminally on erect hyphae coming between epidermal cells of leaf.
5. A layer of basidia with 4-8 basidiospores.

***Dacrymyces* Nees ex. Fr**

1. Basidiocarp gelatinous, cushion shaped or stalked with a spathulate, subglobose or lobed pileus, furcate (tuning forklike).

***Serpula* Pers. Ex. Gray.**

1. *S. lacrymans* (syn. *Merulius lacrymans*), a house fungus.
2. In damp, poorly ventilated houses, causes dry or brown wood rot.
3. Beads of water can be seen on the rotting wood hence, weeping (Lacrymans)
4. Basidiocarp flat, brown coloured, fleshy with pitted hymenophores on lower surface.
5. Basidiospores rusty and cyanophilous.

***Cantharellus* Singer**

1. Basidiocarp fleshy, funnel shaped.
2. Hymenophore with gill like folds, which are irregularly branched, several species are edible.
3. *C.cibarius*, (the famous European cantharell).
4. Sells costliest in French markets.

***Corticium* Quel**

1. Fruit body thin, flat on substratum looking like a splash of white paint.
2. It is just a thin layer of basidia bearing hyphae, the hymenium is fully exposed, saprophytic or parasitic on wood.

***Stereum* Pers. Ex. Gray.**

1. Basidiocarp small, semicircular in several tiers on the dead wood.
2. Upper surface with several soft hairs, grey to brown.
3. *S.hirsutum* (heart rot of Oak).
4. Basidiospores amyloid).

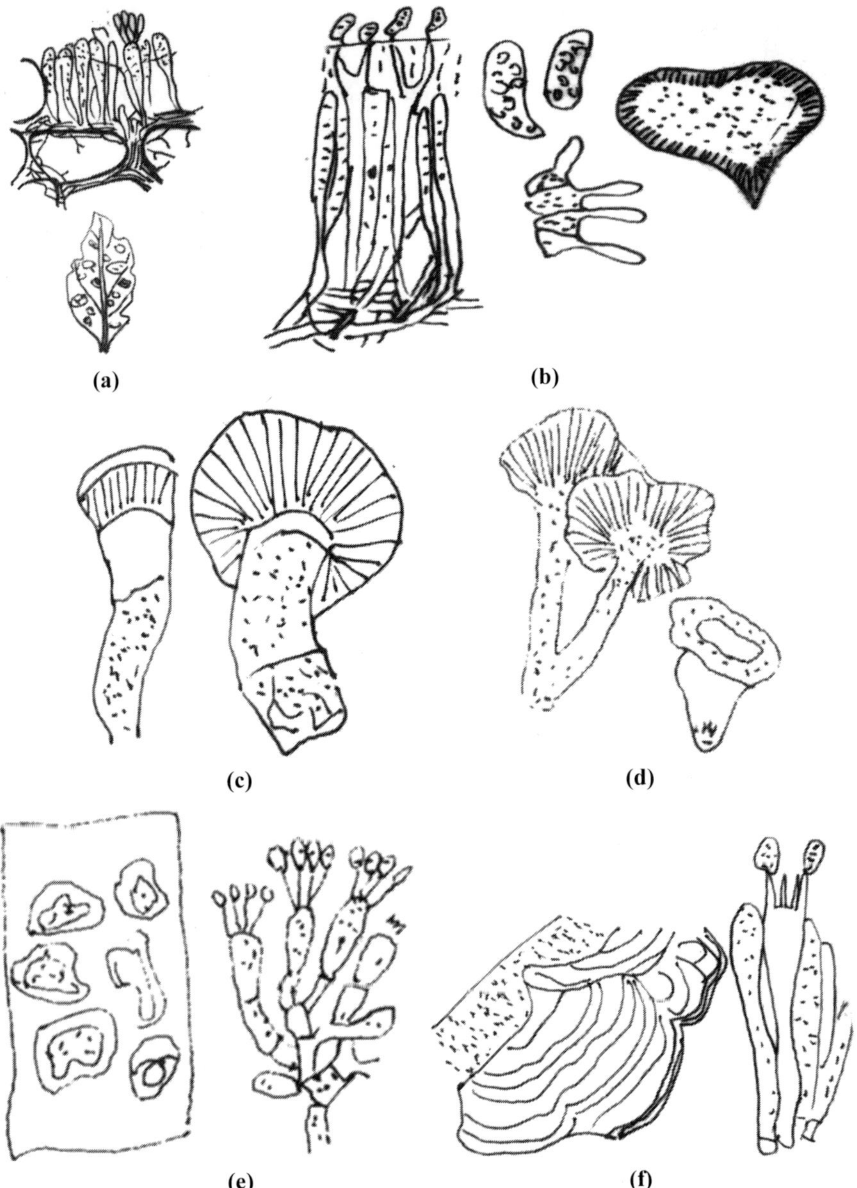

Fig. 6.27. **(a)** *Exobasidium* **(b)** *Dacrymyces* **(c)** *Serpula* **(d)** *Cantharellus* **(e)** *Corticium* **(f)** *Stereum*

***Schizophyllum* Fr**

1. Basidiocarps fan shaped, 1-3 cm dia, hairs on upper surface, gills on lower surface, gills are just longitudinal splits (Schizo-split, phylum leaf of gill).
2. Dry basidiocarps are viable for several years and on moistening regain the shape.
3. Active wood destroyers.

***Clavaria* Vaill ex. Fr. (coral fungus)**

1. Grow in clusters in grass field, on the ground or on wood.
2. Basidiocarps spindle shaped with pointed tips, bright coloured, 5-10 cm high, flattened on groove, adhering closely at the base.
3. Spores white but yellowish in mass.
4. Hymenium without cystidia and cover whole surface.

***Hydnum* L ex Fr.**

1. *A lignicolous* fungus growing on humus, on wood.
2. *H.repandum*, the only species reported from India is edible.
3. Basidiocarp palate, mushroom shaped.
4. The cap is fleshy, smooth, or slightly scaly, pinkish buff in colour, 5-8 cm diameter.

***Ganoderma* Karst.**

1. Cause white rot of wood.
2. *C.versicolour* is common on logs, old stumps, and branches of trees.
3. Basidiocarp semicircular or kidney- shaped, thin, velvety, 2-5 cms with zonation of different colours.
4. Poroid hymenophore.

***Fomes* Fr.**

1. On logs, stumps, and trunks of trees.
2. Perennial, sessile, with extremely minute pores.
3. Basidiocarp sessile, leaf-shaped having extremely minute pores.
4. *F. tomentarius* causes heart rot in conifers and hard wood.

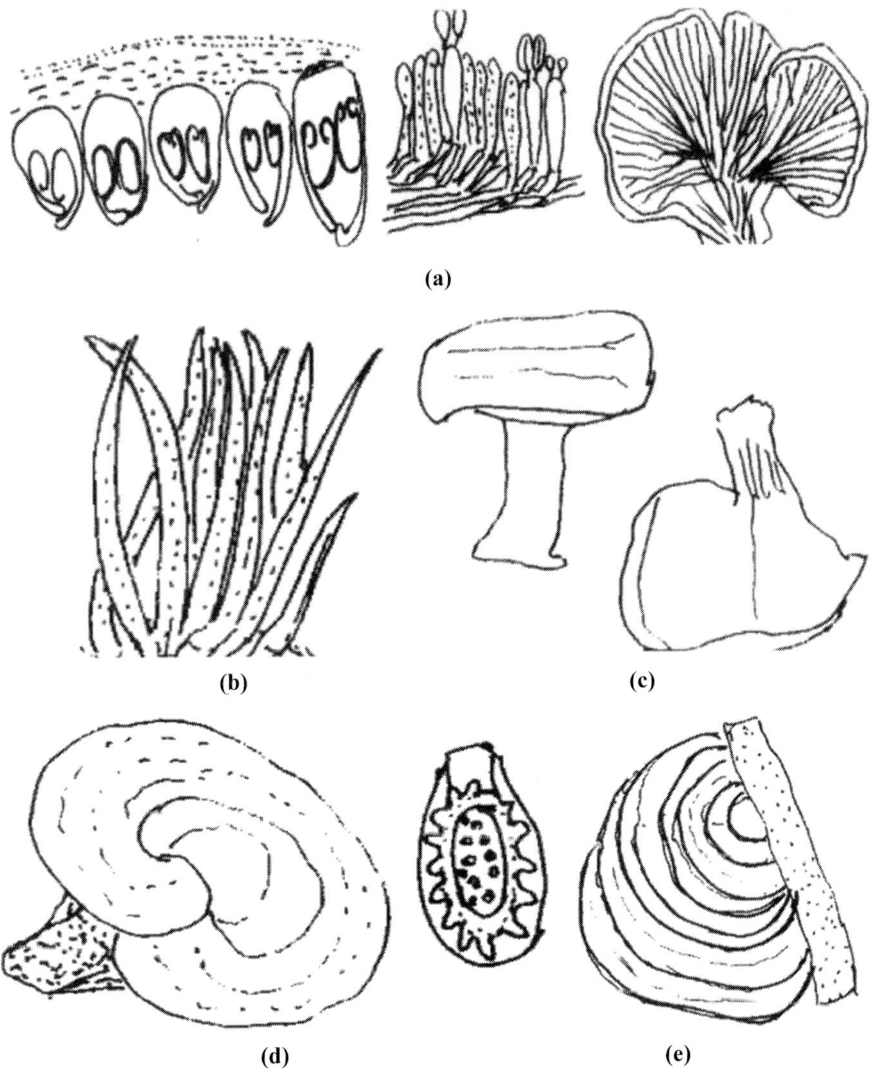

Fig. 6.28. **(a)** *Schizophyllum* **(b)** *Clavaria* **(c)** *Hydnum* **(d)** *Ganoderma* **(e)** *Fomes*

Polyporus Mich. Ex. Fr.

1. Highly destructive wood rotting parasites of trees.
2. Basidiocarps annual, bracket or fan-shaped, stipitate, stalked or laterally attached to the stem. The hyphae in the form of sclerotium, survive inside the tree and form the fruit bodies again in the next year.

3. *P.sulphureus, P. squamosus, P.batutinas* and *P.glomeratus* form fan shaped fruiting, attached laterally to tree trunks by a cap, made up of dikaryotic, tertiary mycelium.
4. The pileus fleshy, convex to flaty, whitish, red, yellow, purple, orange, pink, green or brown.
5. Gills on lower side of the pileus, from periphery and coverage towards the stipe.
6. Hymenium consists of elongated, clubshaped basidia with basidiospores.

***Hericium* Gray.**

1. A lignicolous fungus causing white rot of coniferous and anginospermic timber.
2. Basidiocarp coralloid, beautiful with positively geotrophic spines, attached to the substratum.

***Lenzites* Fr.**

1. Fruit body annual, with lamellate hymenophore, spores hyaline, cylindric, smooth, cystidia fusoid.
2. Pore tube are of different length.

***Armillaria* (Fr) Staude**

1. Honey or Boot lace fungus, (*A. mellea*) popularly known as honey fungus' due to taning coloured basidiocarps and as 'boot lace fungus' due to black coloured rhizomorphs.
2. It is a root rot parasite.
3. It spreads in soil, infect root of trees with shoestring like rhizomorphs.
4. Basidiocarp small, stalked, umbrella like. Annulus present, gills adnate.
5. Basidiospores white, no role in the infection and the spread.

***Pleurotus* (Fr.) Quel.**

1. Oyster fungus, edible mushroom, basidiocarps with stipe, lateral, excentric or even absent.
2. Grows naturally on the trunks and stumps in shelf, like layers.
3. It can be grown as mushroom cultivation.
4. The cap (pileus) deep grey soon turns pale and brownish.
5. The surface is smooth, moist with edge, thin downwards.

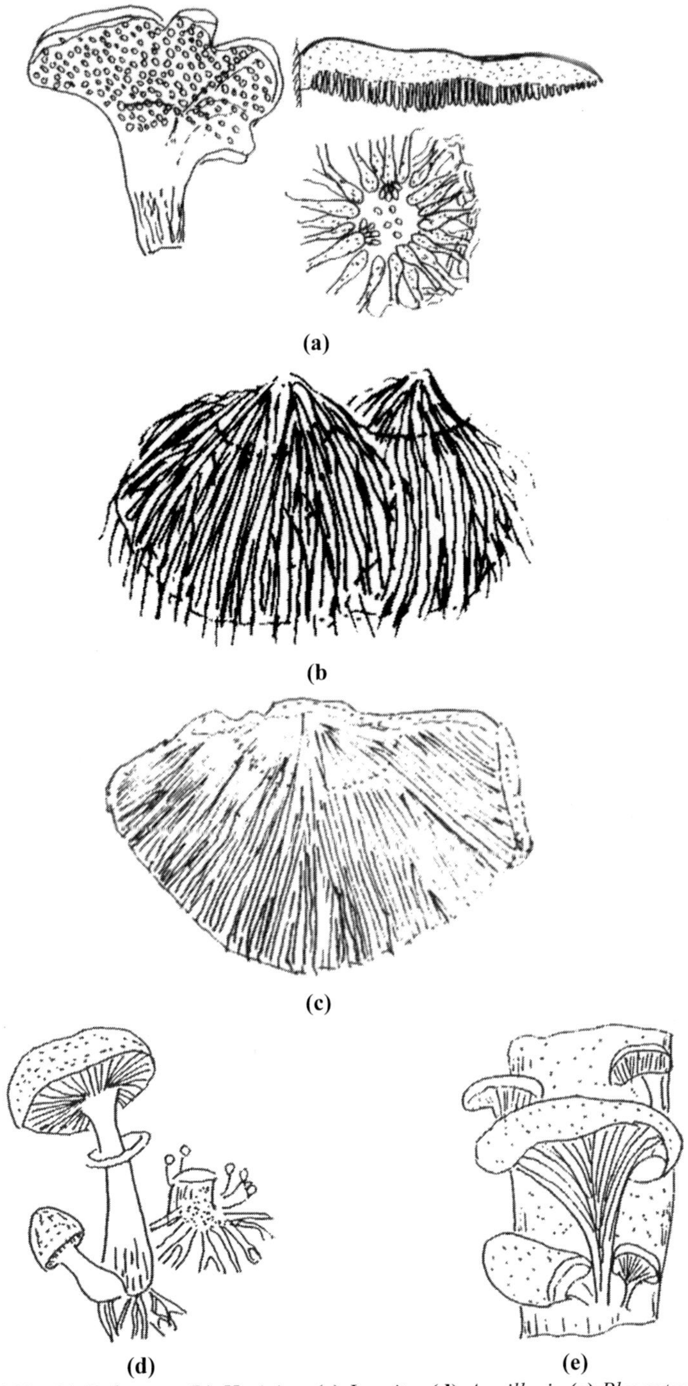

Fig. 6.29. (a) *Polyporus* **(b)** *Hericium* **(c)** *Lenzites* **(d)** *Armillaria* **(e)** *Pleurotus*

***Amanita* Pers. Ex. Hook.**

1. *Amanita verpa, A. phalloides, A.muscaria* are deadly poisonus, known as death caps.
2. The basidiocarps are stalked and umbrella type.
3. Presence of volva and annulus.
4. Pileus with scales on upper surface and hymenium with highly developed free gills toward lower surface.
5. Basidiospores white.
6. *A.muscaria* known as fly fungus for its use in killing flies *A.virosa* (death angel) very beautiful, poisonous.

***Agaricus* Fr.**

1. Button mushroom, Basidiocarp stiped, annulus present, volva absent, edible.
2. *A campestris* and *A. bisporus* (2-spored basidia) cultivated on wheat straw compost on large scale.
3. Gills deep purplish brown, free.

***Psilocybe* Fr Quel.**

1. The sacred fungus, basidiocarp umbrella shaped, contain hallucinogenic substances.
2. *P.semilanceata* commonly known as liberty cap on account of the conical and pointed cap of its fruit body, the symbol of French republic.
3. The hallucinogenic property has been due to two chemicals present named psilocybin and psilocin.

***Coprinus*: Gray**

1. The inky cap grows on humus rich soil, compost heaps and herbivore dung.
2. *C. comatus* is edible.
3. The basidiocarp initially elongated barell shaped cap with white scales on all over the surface (hence the name shaggy cap.)
4. Matured basidiocarps are bell- shaped, pinkish to black.
5. The stipe white, smooth and with annulus.
6. The caps are edible only in young stage and before their autodigestion, the property of self melting of mature lamellae into inky fluid.
7. The whole cap is soon converted into a black liquid.
8. Hence the name 'inky caps'

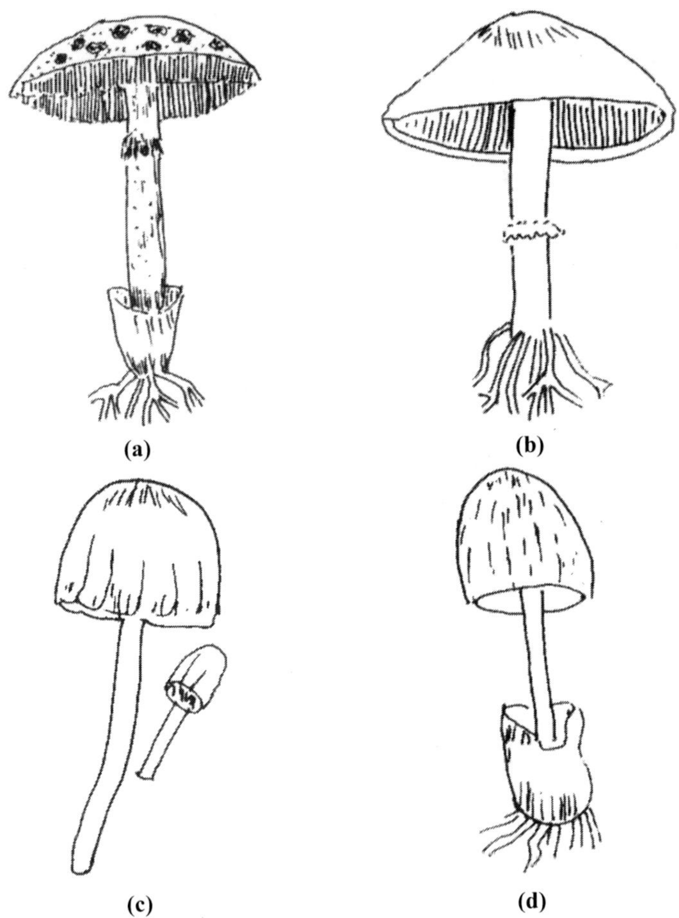

Fig. 6.30. (a) *Amanita* **(b)** *Agaricus* **(c)** *Psilocybe* **(d)** *Coprinus*

Lactarius Gray

1. Miky mushroom, basidiocarp found abundant after rains in the association of forest trees.
2. The cap 5-12 cm in diameter, fleshy and centrally stalked, initially convex but on maturity flatten and finally turn to funnel shaped with the margin turned downward.
3. The cap surface is orange or carrot- red with concentric zones of deeper colour.
4. A milky latex flow out of the fruit on its cutting or wounding, hence the name ‘milky fungus; The latex is coloured in some cases.
5. *L.deliciosus*, the saffron milk- cap, an excellent edible mushroom.

***Pholiota* (Fr.) Kummer,**

1. Basidiocarp in tufts at the base of tree, umbrella-shaped, yellow to light tan.
2. Dark recurved scales present all over the pileus and on stipe.
3. Basidiospores brown.

***Mycena* Gray.**

1. Basidiocarp small, delicate, conical bell- shaped.
2. Pileus papery thin, stipe slender.
3. Growing in clusters all the year round on old stumps, lawns etc.

***Volvariella* Speg.**

1. Grow during rainy season under shades.
2. Basidiocarp with stipes without ring, enclosed in a large cup shaped volva.
3. The pileus is fleshy white or pigmented, edible, spores pink.
4. *V. volvacea* and *V.diplasia* cultivated as paddy straw mushroom.

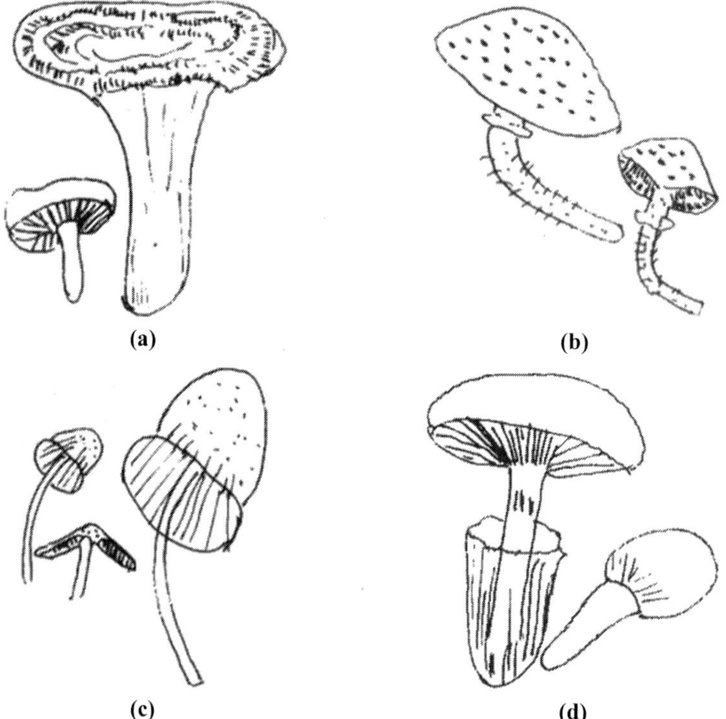

Fig. 6.31. (a) *Lactarius* **(b)** *Pholiota* **(c)** *Mycena* **(d)** *Volvariella*

6.5.3 Class – Gasteromycetes

***Scleroderma* Pers.**

1. Basidiocarp sessile, round, and flat, 5-8 cm in diameter, grow most often on bare soils, in wood and waste lands.
2. The peridium thick, yellow, hard, rough, powdery, and purplish black without any capillitium.
3. Spores with reticulate or spiny ornamentation, get released on irregular rupture of the peridium.

***Lycoperdon* Taurnef.ex Pers. (the puffball)**

1. The basidiocarp pear shaped or globose, 2.5 cm in diameter.
2. The peridium two layered, exoperidium fragile and covered with fragile, cone-like spines.
3. On rupturing the exoperidium, the endoperidium exposed.
4. The mature gleba (fertile portion) consists numerous irregular chambers lined with basidia bearing spherical basidiospores.
5. On maturity the spore mass dry like a face powder.
6. During spore dispersal, visible rings of spores mass are seen, which resemble to smoke rings blown out by an expert smoker.

***Geastrum* Pers. (The earth star)**

1. The onion-shaped basidiocarp grow on leaf litter.
2. The exoperidium 3 layered, the inner most is fleshy, ruptures along radial lines from apex to downward and ray like lobes are formed.
3. The endoperidium form as apical pore, through which spores are released in batches.

***Phallus* Hadr. Jun. ex. Pers. (stink horn)**

1. The mycelium burried in wood, from which rhizomorphs originate, spread to a long distance in soil and then form the basidiocarp.
2. The young fruit body resemble to egg but slightly enlarge and soft inside, which 'stink horn stage' (mature fruit body) is shaped with a short stalk.
3. Peridium three layered, exo- and endo-peridia are papery, mesoperidium thick and gelatinous.
4. The stipe with spongy tissue, surrounded by bell or thimble-shaped pileus, which is attached at apex and free below.
5. At maturity, the stipe elongates pushing the peridium, carrying pileus and gleba at its apex.
6. Soon then gleba undergoes autodigestion, trama and basidiospores lie suspended in slimy mass.

7. The gleba produce sugary substance which are irritable in smell (stinking), but to attract a particular type of insect.

Cyathus **Haller ex. Pers. (The bird nest fungus)**

1. The basidiocarp globose initially but later become funnel shaped, due to rupture of papery epiphrag.
2. The mature fruit body 1.5 cm high and 1 cm in diameter, reddish brown, hairy outside, inner surface shining- grey with longitudinal striation.
3. The peridioles lentil shaped, whitish, 2-3 mm diameter, lie attached with funiculus, neatly nestled inside the fruit body.
4. The funiculus made of three portions, the sheath, middle piece, and purse.
5. The purse is closed below, packed with highly coiled, elastic, thread-like structure.
6. The funicular cord; whose lower free terminal sticky end, known as hapterion.
7. During dissemination of peridioles with the help of rain drops, the fruit body acts like a splash cup.
8. On detachment of peridioles, the sticky hapteron gets stuck to tree branch or leaf and hangs down by stretched funicular cord.

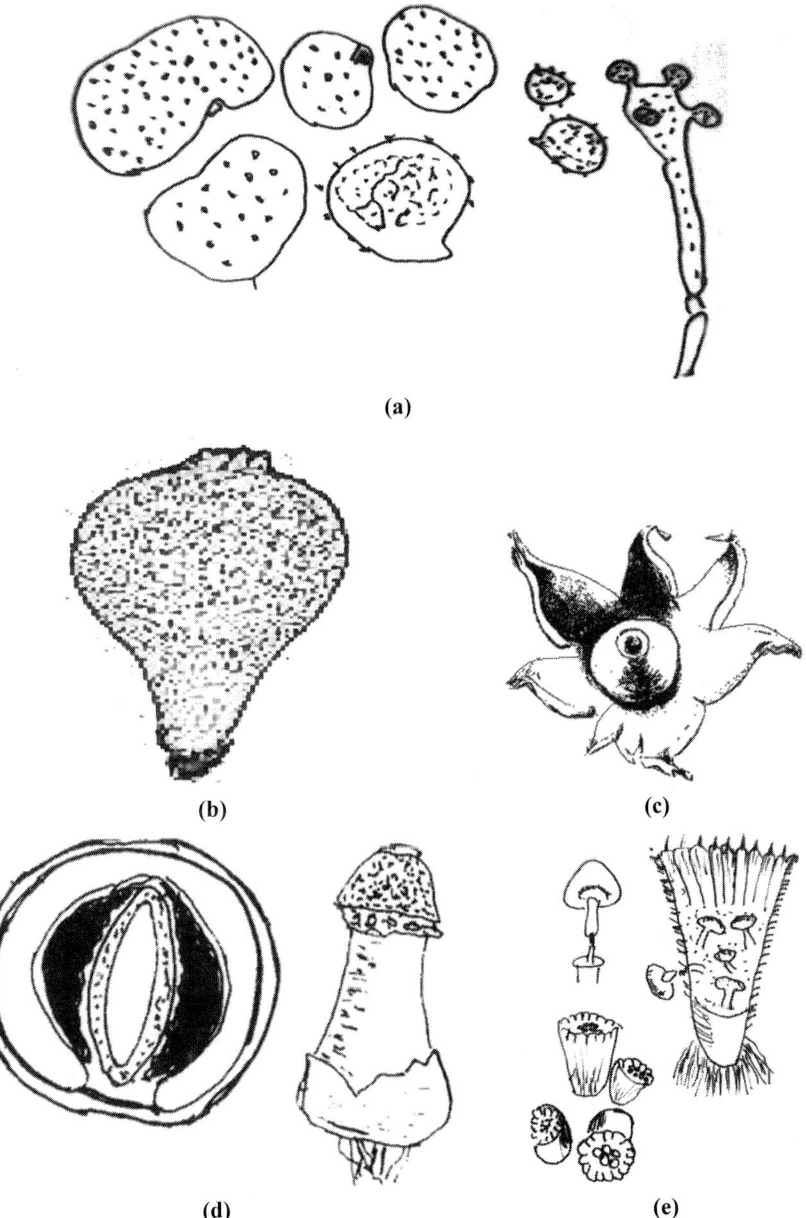

Fig. 6.32. **(a)** *Scleroderma* **(a)** *Lycoperdon* **(c)** *Geastrum* **(d)** *Phallus* **(e)** *Cyathus*

6.6 Sub Division (V) Deuteromycotina

6.6.1 Class- Hyphomycetes

***Aspergillus* link**

1. Some species have perfect state from *Eurotium*, *Sartorya*, *Emericella* and Hemicarpentales.
2. *A. flavus* is known to produce aflatoxins, and *A. niger*, good source of citric acid.
3. Mycelium, septate, branched, hyaline, colonies green, dark green, black, yellowish etc.
4. The conidiophore arises from a special cell (foot cell) of mycelium.
5. The erect hypha forms globose swelling, the vesicle at the tip.
6. On the vesicle 1-2 rows of bottle-shaped phailides (sterigatae) are formed, which end in the chains of spherical conidia.
7. Conidia vary in colour, shape and size.
8. Cleistothecia found in some species.

***Penicillium* Link**

1. Vegetative hyphae creeping, septate, branched.
2. Conidiophores erect, usually unbranched, septate, at the apex vertical of erect primary branches.
3. Each with a vertical of secondary metulae or sometimes with tertiary branchlets, or with vertical of conidial chain bearing phialids, ultimately forming a typical brush like structure.
4. Food cell absent. Conidia coloured, globose ovate or elliptical, smooth, or rough.
5. Colonies grey-green, olive-green, deep-green, bluish-green etc.

***Scopulariopsis* Bainier Thom**

1. Conidiophores very short. Sometimes single phialides with chain of spores, scattered on mycelium.
2. Conidia pointed at apex and truncate at base with thickened basal ring surrounding a basal germinal pore.

***Gliocladium* Corda**

1. Conidiophores erect, simple or branched, septate, producing at the apex composed fructification due to successive verticils of primary, secondary branches, metulae and phialides.
2. The conidial head is enveloped in slime, conidia in chains.
3. Colonies on agar broadly spreading, floccose, surface white to pale cream.

4. *G.roseum, G.fimbriatum* and *G. atrum* common in soil.

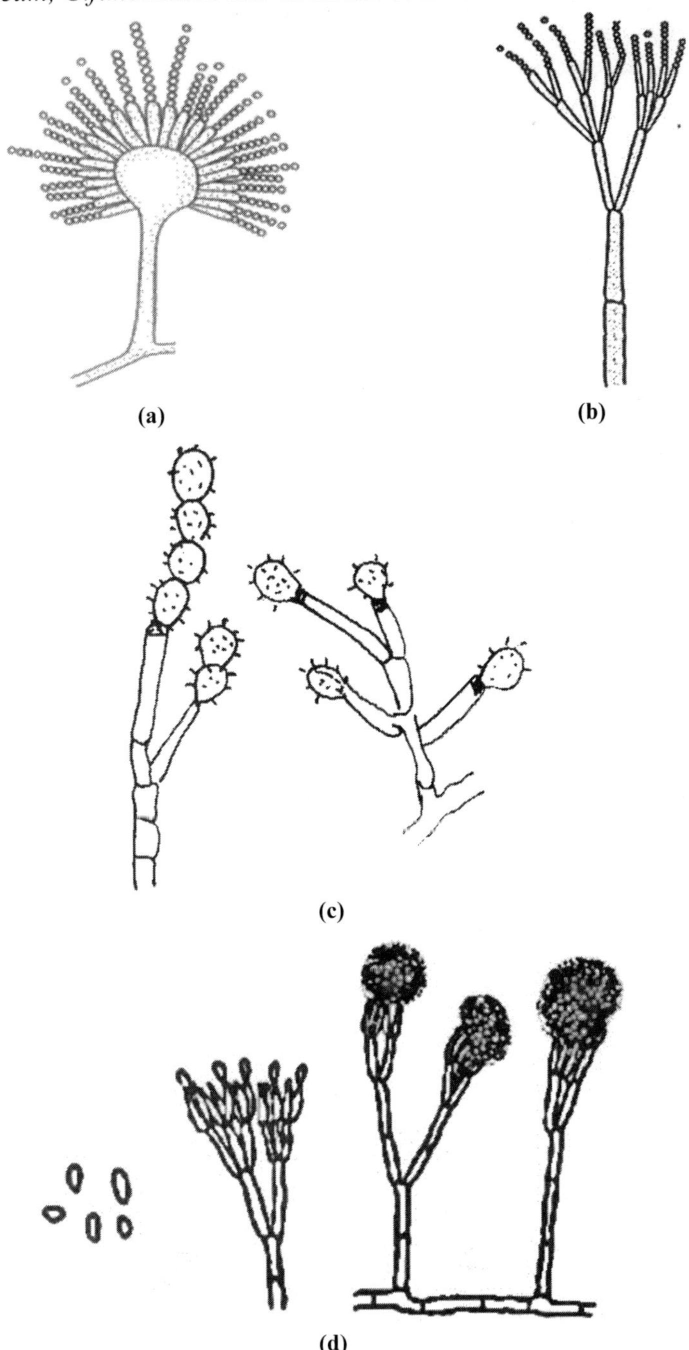

Fig. 6.33. (a) *Aspergillus* **(b)** *Penicillium* **(c)** *Scopulariopsis* **(d)** *Gliocladium*

Acremonium **Link**

1. Hypha forming furg, branched, septate, prostate possessing side branches, which become erect and serve conidiophores.
2. Conidia single on conidiophores, terminal, hyaline or bright coloured, small ovate.

Spicaria **Harting**

1. Conidiophores erect, septate, usually freely branched, often in whorls, or irregular, each branchlet with a terminal fructification, composed of a verticil of divergent metulae, on which verticil of divergent phailides.
2. Heads divergent and look penicillate, conidial chains usually long, conidia hyaline, round or elongate.
3. *S.griseola, S.elegans, S.violacea* are common in soil.

Verticillium **Nees**

1. A whorl of branches.
2. Mycelium septate, hyaline to light coloured.
3. Conidiophores, erect, septate, branched. Branches of the first order are whorled, opposite or alternate. Branches of second order whorled, dichotomous or trichotomous on the branches of first order;
4. Terminal branchlets usually flask –shaped pointed at tip.
5. Conidia always borne single on the branchlets, round, elliptical, inverted egg shaped hyaline or light coloured.
6. *V. albo- atrum, V. glaucum, V. effusum* are commom in soil.

Paecilomyces **Bainer**

1. Conidiophore branches more divergent than *Penicilium*.
2. Conidia dry in basipetal chain, one celled, ovoid to fusoid, hyaline saprobic.

Trichoderma **Pers**

1. Mycelium rapid growing, whitish yellow, green to bright green.
2. Conidiophores, erect, septate, unbrached.
3. Conidia (Phialospores) single celled, hyaline, or green.
4. *T.viride* is common in soil- borne plant pathogens.
5. Useful in biological control of soilborne plant pathogens.

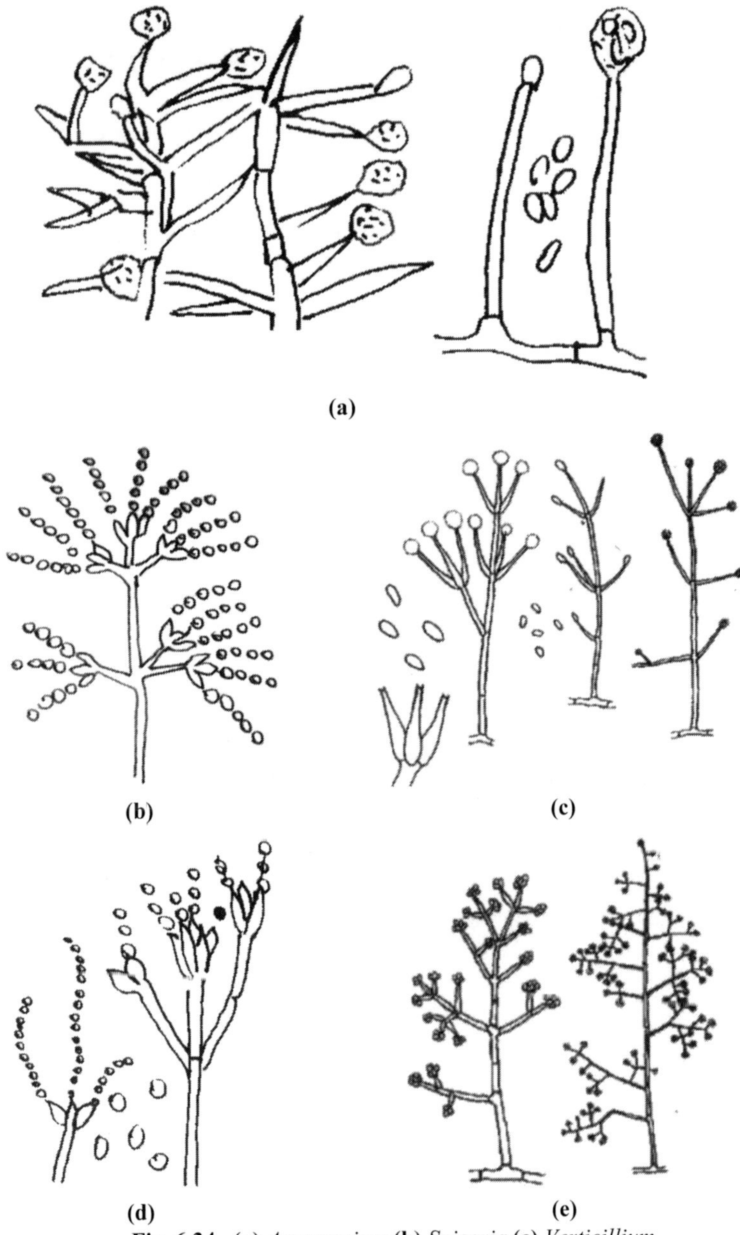

Fig. 6.34. (a) *Acremonium* **(b)** *Spicaria* **(c)** *Verticillium* **(d)** *Paecilomyces* **(e)** *Trichoderma*

***Monilia* Pers.**

1. Conidiophores branched, arise laterally or terminally from the hyphae.
2. Conidia bead like, oval or lemon shaped in long, branched chains.
3. The imperfect stage for several species belonging to Ascomycotina.

***Botrytis* Micheli**

1. Conidia look like grape bunches on conidiophores, which are brown, erect, simple on short sterigmate, unicellular, oval.
2. In culture irregular sclerotia are seen at late.
3. *B. cinerea* (grey mold) is conidial stage of *Sclerotinia*,
4. Colonies, diffuse grey green, olive green, reddish green or brown, black.
5. Conidia dry, loose or dense.

***Trichothecium* Fries**

1. Conidiophores erect, simple, rarely septate, hyaline, cluster of conidia at the apex.
2. Conida hyaline or bright coloured, 2-celled, ovate, nipple like projection at the point of attachment.
3. *T.roseum* is common in soil.

***Lemonniera* De wild.**

1. An aquatic hyphomycete found on decaying leaves in water.
2. Hyphae, hyaline, branched, septate, conidiophore branched, apically and bears phialids.
3. Conidium starts as a small spherical swelling which soon become tetrahedral, long septate armed.

***Cephalosporium* Corda.**

1. Conidiophores and phialides slender, simple, conidia hyaline, 1-celled, collecting in a slime drop.
2. Saprobic or parasitic, some species causes vascular wilts of tree plants.

***Arthrobotrys* Corda**

1. Conidiophores long, simple septate, hyaline, slightly broader at apex, and spore bearing regions.
2. Conidia hyaline, usually 2-celled, born on peg-like denticles, loose clusters.
3. Saprophytic or parasitic on nematodes.

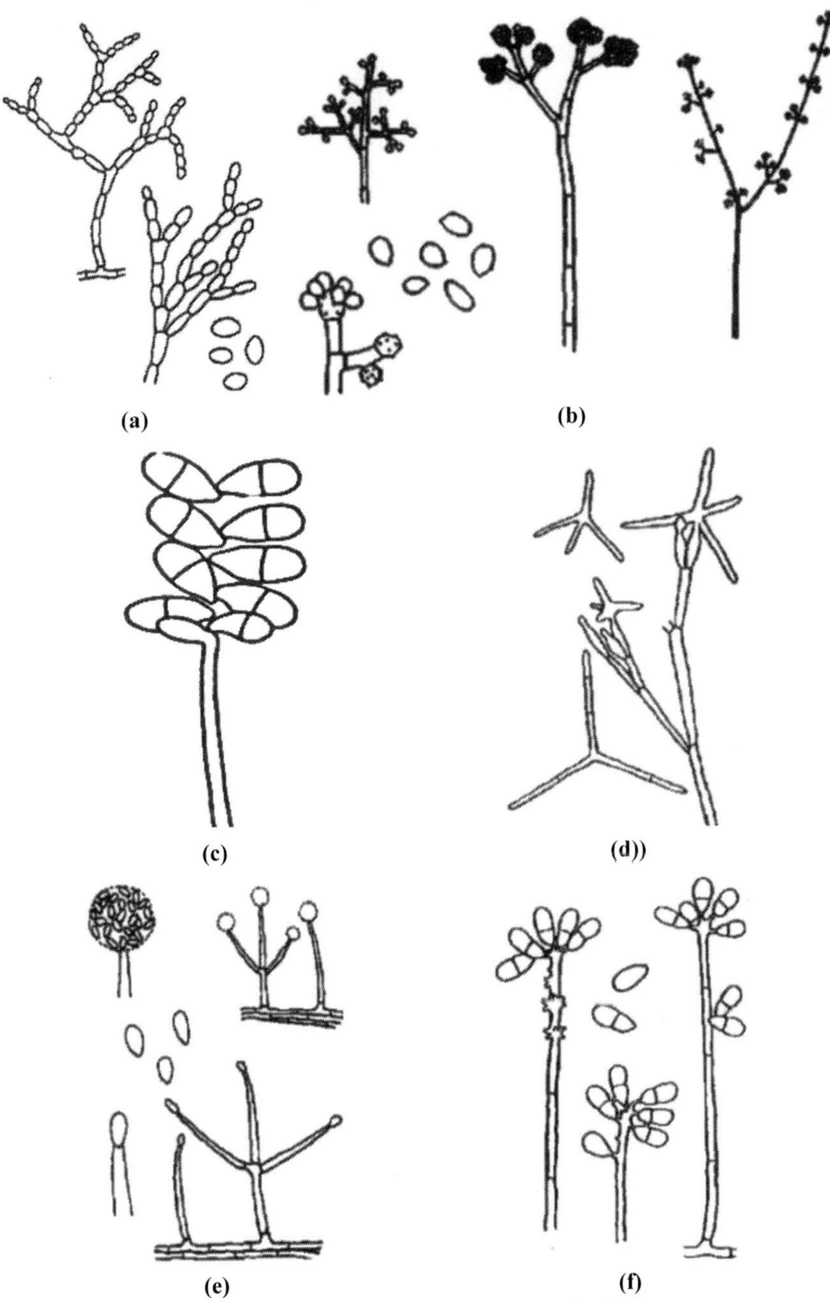

Fig. 6.35. (a) *Monilia* **(e)** *Botrytis* **(c)** *Trichothecium* **(d)** *Lemonniera* **(e)** *Cephalosporium* **(f)** *Arthrobotrys*

Humicola **Traae**

1. Condiophores simple or rarely with short branches, dark.
2. Conidia (aleuriospores) single, apical, globose, brown, 1-celled or phialosporic chain in some cases. Saprobic.

Mycogone **Link**

1. Mycelium hyaline initially, become brown. Conidiophores short, lateral.
2. Conidia single on the tips of the conidiophores, dissimilar two-celled.
3. The upper cell larger, warty, bright coloured. The lower cell pale.
4. *M.nigra*, *M.alba* common in soil.

Cylindrocarpon **Wollen**

1. Imperfect stage of *Nectria.*
2. Conidiophores erect, hyaline, simple or branched terminating in phialides with conidia 3-4 celled, hyaline- in a small fascicle, resemble with *Fusarium* but conidia are not curved and large in size.

Thielaviopsis **Weth**

1. Conidiophores, indistinguished, phialides slightly inflated below, tapering to a long, cylindrical apex.
2. Forming thick-walled very dark aleuriospores. Parasitic or saprobic.
3. Phialospores endoconidia produced inside sporogenous cell in basipetal succession.
4. Conidia variable in length and width, cylindrical, truncate at both ends. Lightly pigmented.

Chalaropsis **Peyron**

1. Conidiophores pigmented, phialides slender, slightly larger near base and tapering up-ward, producing hyaline, phialosporic conidia. endogenously often in chains.
2. Aleuriospores, dark thick-walled, single or in short chains.
3. Parasitic or sarprobic

Chalara **Corda**

1. Mycelium typically dark, conidiophores dark pigmented or hyaline in culture.
2. The basal portion unicellular or septate, the apical cell (phialide) tapering upward and produce conidia endogenously, conidia hyaline, cylindrical, sometime in chains.

3. Parasitic or saprobic.

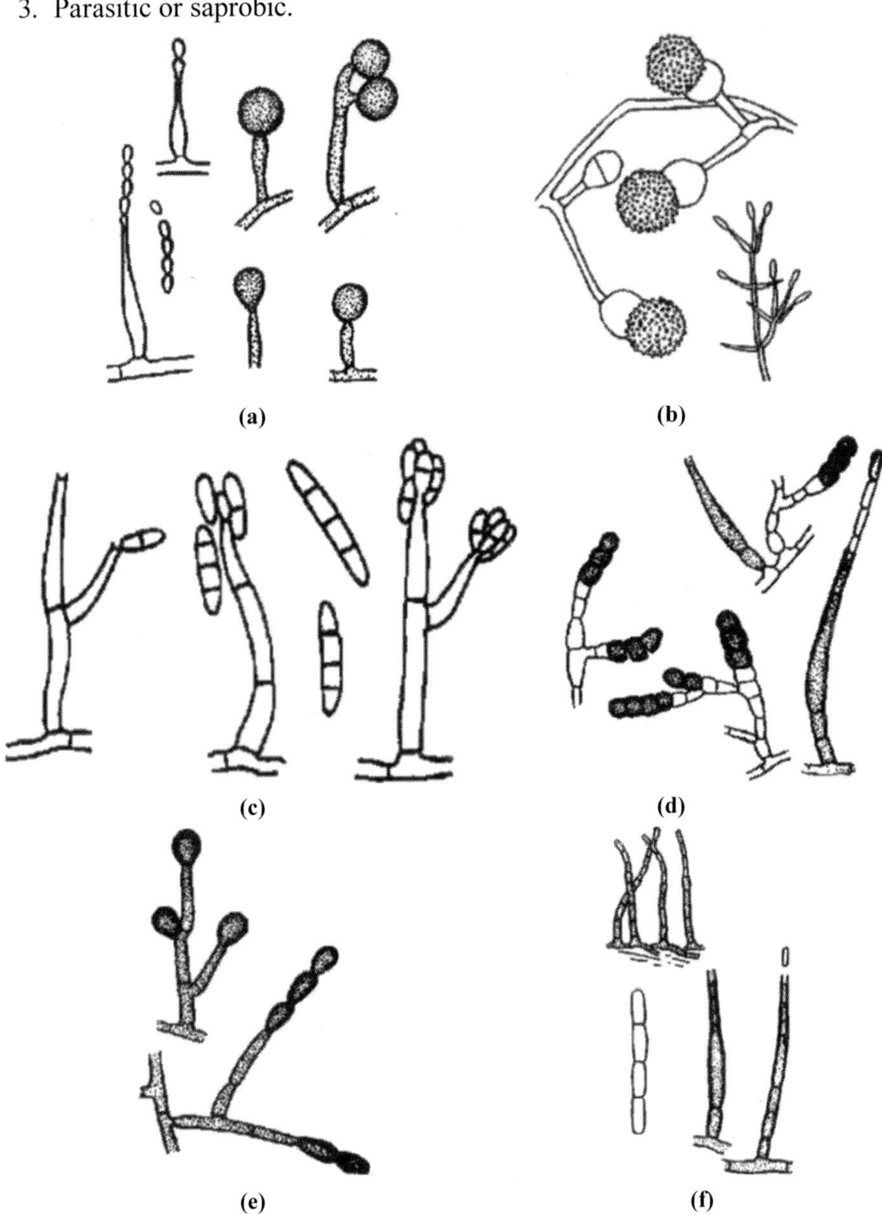

Fig. 6.36. **(a)** *Humicola* **(b)** *Mycogone* **(c)** *Cylindrocarpon* **(d)** *Thielaviopsis* **(e)** *Chalaropsis* **(f)** *Chalara*

***Fusarium* Link**

1. Hyphae hyaline, branched, septate.
2. Conidiophores, hyaline, short, simple, or branched, septate bearing terminal phialids.
3. The phialids are subulate (broad at base, narrow at apex).
4. Conidia slimy are of two types.
5. Microconidia, one celled, oval or comma shaped, pyriform or elongate, hyaline.
6. Macroconidia-spindle shaped, pointed ends, 3 to many septate, with a distinct foot cell.
7. *F.moniliforme, F.lini, F.oxysporum are* common.

***Sphacelia* Lev**

1. Sporodochium stromatic, conidiophore hyaline, simple in palisade layer.
2. Conidia hyaline, small, oval, 1-celled produced in sugary honey dew, conidial stage of *Claviceps*.

***Tubercularia* Tode.**

1. Sporodochia large, light to orange on wood bark.
2. Conidiophores hyaline, elongate repeatedly, irregularly branched, conidia terminal, hyaline, 1-celled.

***Myrothecium* Tode**

1. Sporodochia cushion like, sometimes with marginal hyaline setae.
2. Conidiophores branched, sub-hyaline to coloured.
3. Conidia sub-hyaline to dark, 1 celled, ovoid to elongate. Parasitic or saprobic.

***Ramulispora* Miura**

1. Sporodochia small from sub-stomatal region and coming out through stomata.
2. Conidiophores hyaline, simple or branched, short.
3. Conidia hyaline, filitrom, septate with short lateral branches. Superficial sclerotia present.
4. Parasitic on leavers.

***Epicoccum* Link**

1. Sporodochia and, conidiophores dark, short conidia, dark, dictyosporous, globose.
2. Saprobic or weak parasitic.

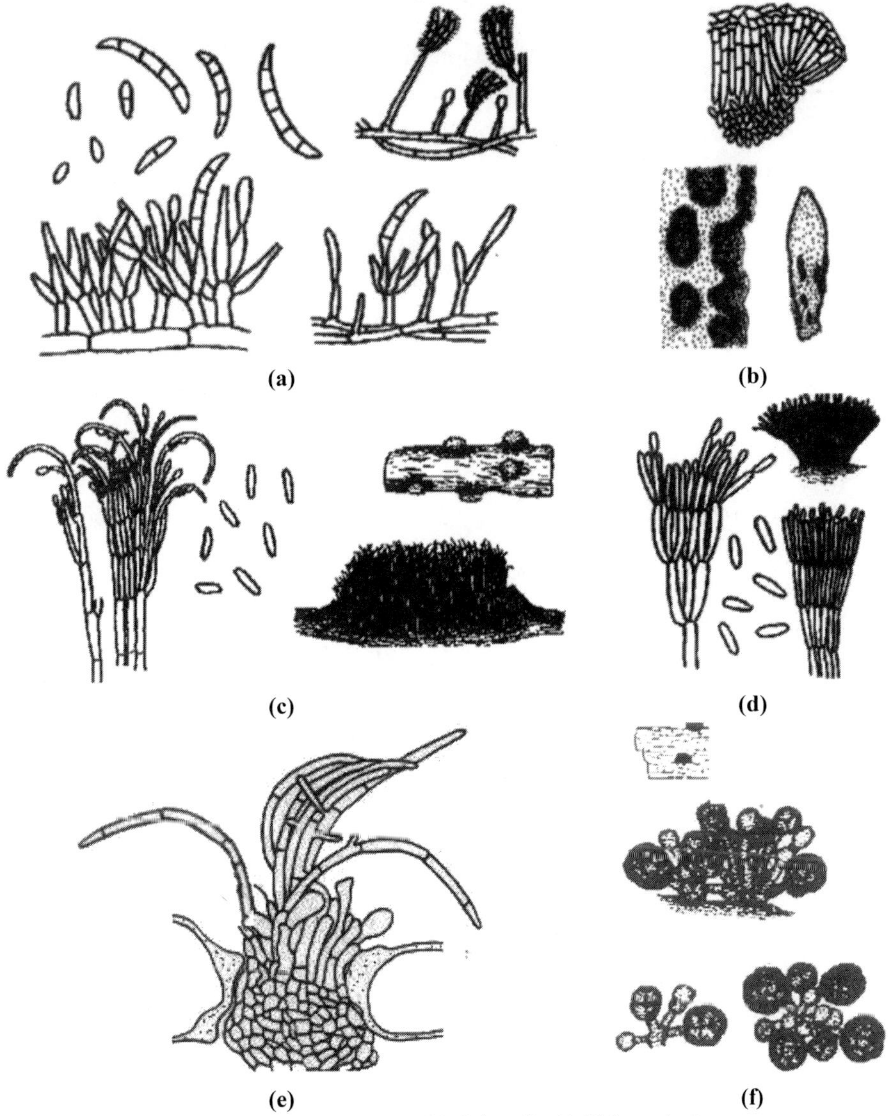

Fig. 6.37. **(a)** *Fusarium* **(b)** *Sphacelia* **(c)** *Tubercularia* **(d)** *Myrothecium* **(e)** *Ramulispora* **(f)** *Epicoccum*

Stysanus Corda

1. Coremia erect, clubbed-cylindric, dark coloured, rigid.
2. Conidia in loose, long or globose panicle, ovate, lemon shaped. Hyaline in chains.
3. *S.medins, S.slemonites* common in soil.

***Trichurus* Clem & Shear**

1. In sporiferous head both sterile and fertile hyphae are mixed.
2. Synnemeta dark with expanded spore bearing portion.
3. Conidiogenous cells monoblastic.
4. Conidia unicellular, ellipsoid, ovoid or sub spherical, truncate at the base, smooth walled, pale brown.

***Graphium* Corda**

1. On wood substrate or plant debris.
2. Synnemeta dark with glistening mass of amerospores in mucus, in annellidic manner.
3. Perfect stage *Ceratocystis ulmi.*

***Stilbella* Lindau**

1. 1. Coremia with definite stipe and definite head; hyaline or bright coloured.
2. Stipes with parallel branched hyphae, which diverge at apex, forming the head.
3. Finally, branched hyphae serve as conidiphores.
4. Conidia solitary but held togather in a mucous, ovate, elongate, or globose, very small, hyaline.

***Isaria* Pers.**

1. Synnemata light coloured, conida 1 celled, hyaline, ovoid, dry with no gelationous material.
2. Saprobic or parasitic on insects.

***Pyricularia* Sacc**

1. Conidiophores erect, simple, septate and hyaline or grey.
2. Conidia, hyaline or pale grey, bicelled.
3. *P. oryzae* causes blast of rice.

***Ramularia* Sacc**

1. Conidiophores well developed through stomata of host leaves, stout, clustered, hyaline or sub-hyaline, frequently curved with prominent conidial scars.
2. Conidia hyaline, cylindrical, typically three celled but many are one celled and few three celled sometimes in short chains.
3. Causing leaf spots.
4. *R.areola* causes grey midew of cotton.

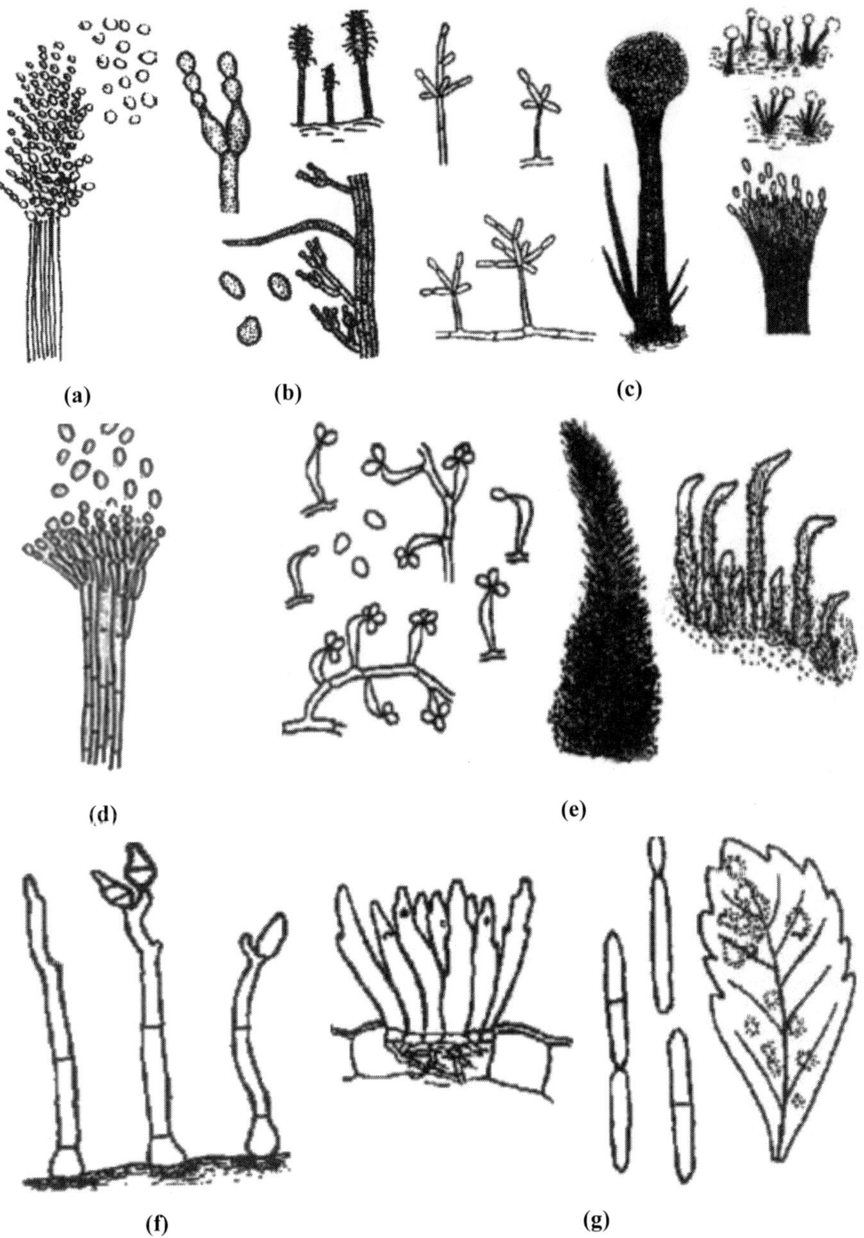

Fig. 6.38. **(a)** *Stysanus* **(b)** *Trichurus* **(c)** *Graphium* **(d)** *Stilbella* **(e)** *Isaria* **(f)** *Pyricularia* **(g)** *Ramularia*

***Helicoma* Corda.**

1. Conidiopores dark, shout, septate.
2. Conidia hyaline or dark, tightly curled, Saprobic.

***Tetraploa* Berk & Vr.**

1. Conidiophores absent.
2. Conidia directly on mycelium, 4-longitudinal cells developed into appendages, brown, smooth or rough, Saprobic.

***Diplocladiella* Arnaud.**

1. Conidiophores erect.
2. Conidia apical or lateral with two pointed and septate arms.
3. Central cell dark, saprobic.

***Helicosporium* Nees.**

1. Conidiophores tall, slender, brown, septate, simple, or branched.
2. Conidia hyaline or pigmented, septate, coiled, saprobic.

***Diplococcium* Grove.**

1. Conidiophores erect or ascending, frequently branched, brown.
2. Conidia (porospores) mostly 2-celled, short brown in acropetal chain.
3. Saprobic on wood bark.

***Dictyosporium* Corda.**

1. Conidiophores dark, slender simple or branched, short bearing a branched conidium apically, sometimes in sporodochium.
2. Conidia with many close septa, branche arising from different points, saprobic.

***Helminthosporium* Link**

1. Helminthos a wormlike, hyphae typically dark, septate, branched.
2. Conidiophores simple, erect, brown, bearing conidia at tip and laterally, often forming whorles above or below septa.
3. Conidia obclavate, brown, thick walled, tranverse septate, 8-10 cells and basal scar of attachment.
4. Parasitic species of *Helminthosporium* are included in two genera *Drechslera* and *Bipolaris*.

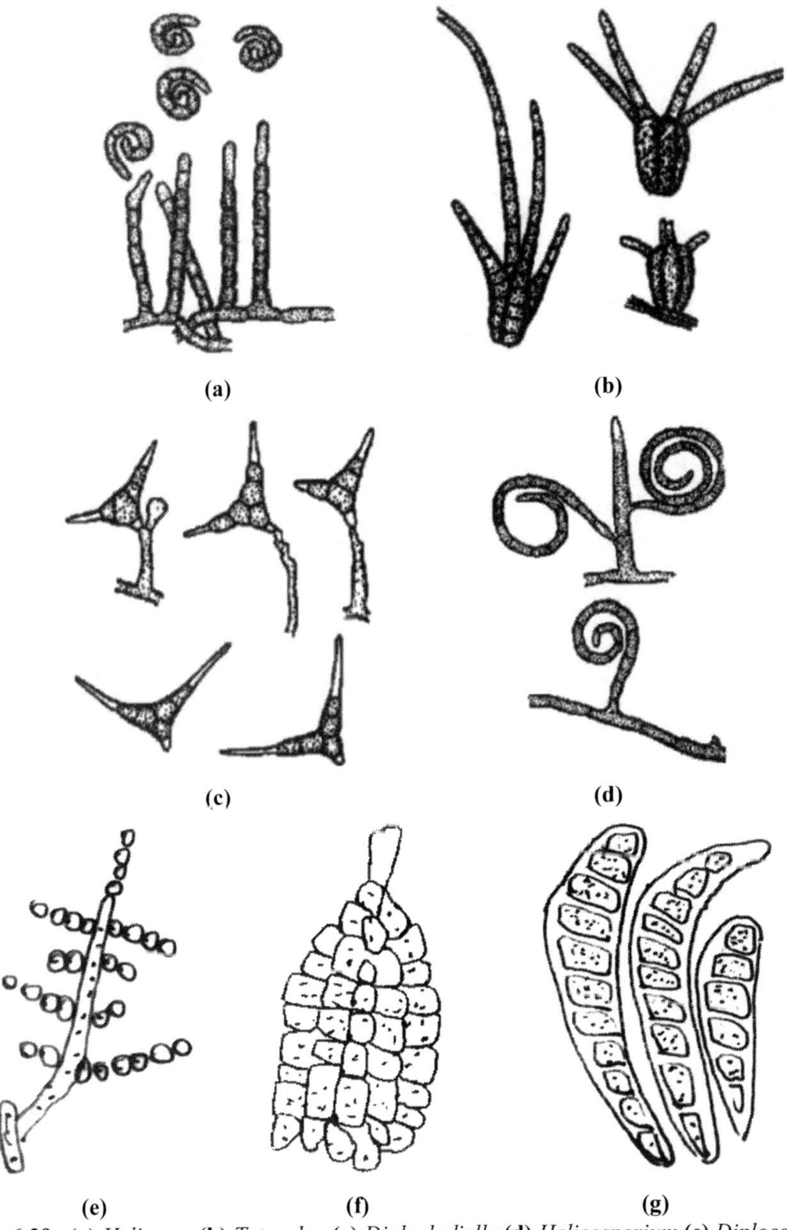

Fig. 6.39. (a) *Helicoma* **(b)** *Tetraploa* **(c)** *Diplocladiella* **(d)** *Helicosporium* **(e)** *Diplococcium* **(f)** *Dictyosporium* **(g)** *Helminthosporium*

***Drechslera* Ito**

1. Cylindrical conidia, multiseptate, acropetally arranged.
2. Conidiophore sympodial, extending.
3. Perfect states *Cochliobolus*, *Pyrenophora, Pleospora* and *Trichometasphaera.*
4. *D.oryzae* pathogenic on rice.

***Bipolaris* Shoemaker**

1. Conidiophore brown, simple producing conidia through apical pore.
2. Conidia (prosporous) brown, several celled, elliptical, straight or curved, germinating by one germtube at each end (bipolar).
3. Parasitic mostly on grasses.

***Curvularia* (Boedijn)**

1. Cresent or half moon shaped, conidiophores brown, septate, unbranched, geniculated, sympodula bearing cluster of conidia.
2. Conidia dark brown, 2-3 septate, mostly curved at third and large cell from the base.
3. *C. lunata, C.geniculata, C. tetramera, C. interseminata* (spores tapering both ends).

***Alternaria* Nees**

1. Conidiophores dark, broader than hyphae, sympodula, bearing geniculations (scars of deatched conidia).
2. Conidia muriform (both tranverse and longitudinal septa), beaked.
3. Common species *A.tenuis, A.brassicae, A.brassicicola* and *A.raphani.*

***Stemphylium* Wallroth**

1. Hyphae dark coloured, septate.
2. Conidiophores arise as side branches, short, mostly unbranched.
3. Conidia singly, terminaly, ovate or club shaped, muriform, dark coloured.
4. *S.botryosum, S.piriforme* common in soil.

***Ulocladium* Preuss**

1. Conidiophores determinate, sympodial, dark, simple, septate.
2. Conidia (porospores) dark, dictyosporus, without constrictions at septum, borne singly, apical.

***Periconia* Bon.**

1. Parasitic or saprobic, conidiophores dark, tall, upright, stout, simple somewhat swollen at apex, with a loose head of conidia.
2. Conidia dark, 1-celled, globose, in dry chains.

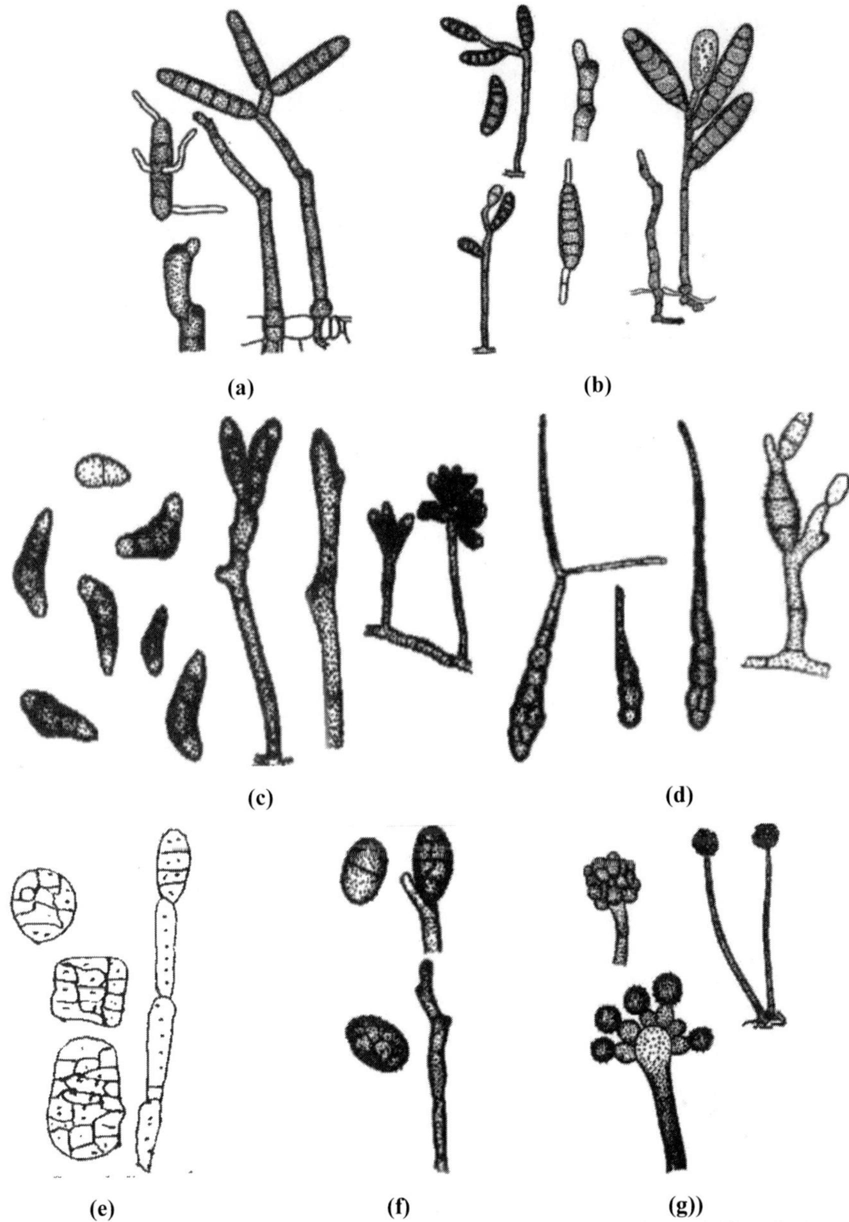

Fig.6.40. Conidia and conidiophore of **(a)** *Drechslera* **(b)** *Bipolaris* **(c)** *Curvularia* **(d)** *Alternaria* **(e)** *Stemphylium* **(f)** *Ulocladium* **(g)** *Periconia*

Pithomyces Berk & Broome.

1. Conidiophores short, simple, peg like, lateral top of the mycelium.
2. Conidia (aleurospores) single apical, several cells (dictyosporous) oblongs to pyriform, verrucose to echinulate. Saprobic.

***Nigrospora* Zimm.**

1. Hyphae septate, branched, hyaline to dark brown.
2. Conidiophores typically short, ampullifrom (flask-shaped) with a single conidium at the tip.

***Cercospora* Fries**

1. Cosmopolitan, weak parasite on dead, dying or diseased plants.
2. Few species pathogenic to man, *Cercospora.personata* and *C.arachidicola* are serious pathogens of groundnut, cause tikka disease.
3. *C.musae* (Sigatoka desease of banana), *C.beticola* (leaf spot of beet) *C.apii* (leaf spot of celery and face lesion in human beings) *C.kikuchii* (purple stain of Soyabean seeds).
4. Conidia, many celled, long, hyaline or pigmented, borne acropetally or aggregate in fascicles. *C.arachidicola* has got perfect stage as *Mycosphaerella arachidicola* and *C.personata* as *M.berkeleyi*. *C.personata* is known as *Cercosporidium personatum*.

***Cercosporella* Sacc.**

1. Conidiophores hyaline, slender, conidia, (sympodulospores), hyaline, several-celled, oblong, cylindrical to filiform, straight or curved, parasitic on higher plants.

***Beltrania* Penzig.**

1. Conidiophores simple or forked, brown.
2. Conidia, (sympodulospores), biconic, 1-celled, brown with a pale middle band.
3. Saprobic *B.indiaca* is common.

Ustilaginoidea

1. *U.virens* causes false or green smut in rice.
2. It grows on ovary, forming greenish-brown clustered mass of hyphae and spores.
3. Spores laterally or terminally on short sterigmata.
4. Spores when young are smooth and round but on maturity become rough, olive green, granular coating.
5. Some of spore-balls develop into sclerotia.
6. Sclerotia on germination produce stroma and a ascomycetous stage.

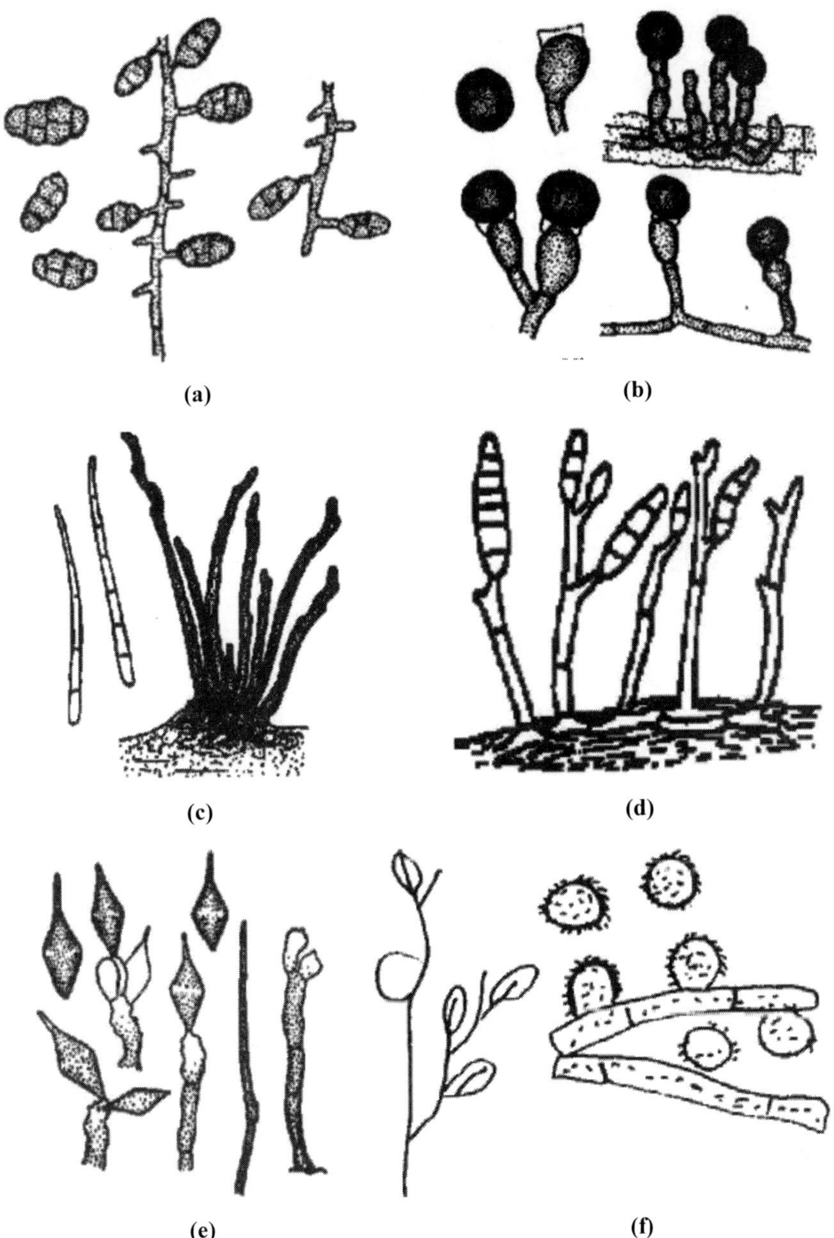

Fig.- 6.41. Conidia and conidiophores of **(a)** *Pithomyces* **(b)** *Nigrospora* **(c)** *Cercospora* **(d)** *Cercosporella* **(e)** *Beltrania* **(f)***Ustilaginoidea*

***Colletotrichum* Corda**

1. Acervulus subcuticular, tan coloured, conidiophores hyaline to light brown, septate, branched at base.
2. Conidia hyaline, unicellular, falcate, or lunate (sickle-shaped), guttulate. On germination conidia become pale brown, septate and form appressoria
3. Species of *Colletotrichum* cause 'anthracnose' (coral-like leaf spots) on several crops.
4. *C.falcatum* (red rot of sugarcane) *C. capsici* (chilli) *C. musae* (banana). *C. graminicola* (maize, jowar) *C. circinans* (onion smudge).

***Gloeosporium*: Desm & mont.**

1. Similar to that of *Colletotrichum* except absence of the setae.
2. Conldia slimy, ellipsoid 1-celled, conidial state of *Glomerella*.

***Pestalotia* de Not:**

1. An acervulus fungus.
2. Conidia 5-celled (the middle three coloured, lower and upper one are hyaline) the upper hyaline bears 2-3 appendages (setulae) while lower hyaline cell has a pedicel
3. Some species cause leaf spot disease.

Pestalotiopsis

1. It differs from *Pestalotia* in having 4 septate conidia with 2 or more apical appendages.
2. Fruiting not acervulus but it is eustromatic and cupulate.
3. *P.theae* (Grey blight of tea).

***Melanconium* Link**

1. Acervulus sub epidermal or sub cortical, black conidiophore simple, conidia dark, 1 to 2 celled ovoids, parasitic or saprobic.

***Monochaetia*: Sacc.**

1. Acervuli dark, discoid, 4 septate condia, with a single apicle, simple or branchd appendage.
2. *M.mali* cause leaf spot apple.

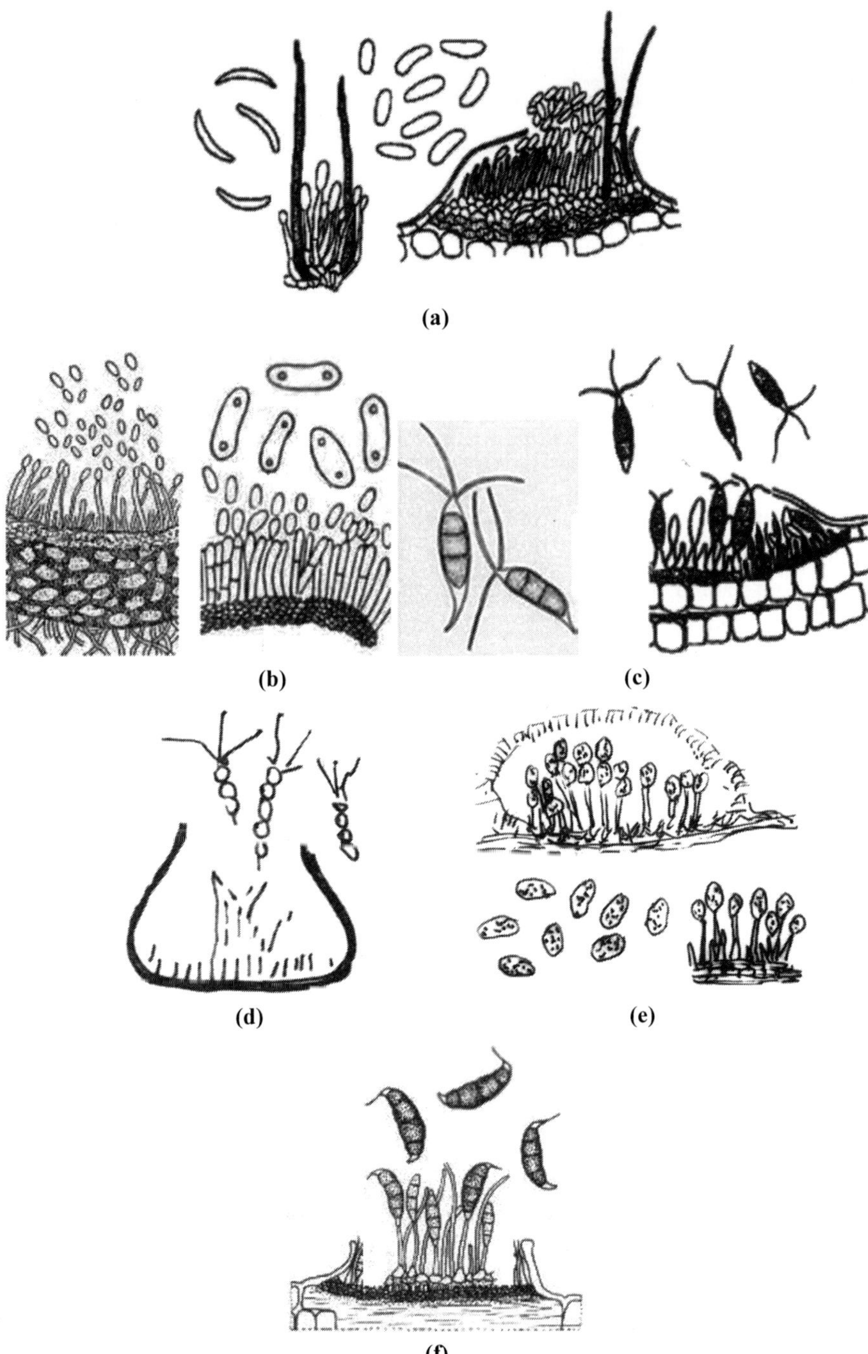

Fig. 6.42. Conidia and conidiophore of **(a)** *Colletotrichum* **(b)** *Gloeosporium* **(c)** *Pestalotia* **(d)** *Pestalotiopsis* **(e)** *Melanconium* **(f)** *Monochaetia*

6.6.2 Class – Coelomycetes

***Phyllosticta* Pers.**

1. Cause leaf spot and shot hole.
2. Pycnidia thin walled, dark brown, uni or multicellular, ostiolate, embedded in subepidermal stroma.
3. The conidiogenous cells are short, cylindrical, or conical, conidia unicellular, hyaline, surrounded by a slime layer and an apical appendage.
4. The perfect stage belongs to *Guignardia* and *Mycosphaerella.*

***Phoma* Desm.**

1. Similar to *Phyllosticta* except that it occurs on dry stem and plant parts than on green leaves.

***Macrophoma* Berl. & Vogl.**

1. *M.phaseoline* (=*Macrophoma phaseolina*) Syn. *Rhizoctonia bataticola*).
2. Pycnidia dark brown, globose or superglobose, ostiolate, solitary or immersed in host tissues become erumpent, dark coloured. Conidiophores (phialides), hyaline short.
3. Conidia hyaline, unicellular, oval.
4. Sclerotia, smooth, black, and hard.

***Phomopsis* Sacc.**

1. Fruit body, (stroma) is thick walled, brown, simple or convulated, ostiolate.
2. Conidiophores branched, septate ending in monophialidic conidiogenetic cells
3. Conidia hyaline and 2-types.
4. L –conidia ovoid to fusoid and B-conidia, stylospores, filiform, curved at the apex like a walking stick with two oil globules.
5. Parasitic, cause spots on plants.

***Ampelomyces* Ces.**

1. Pycnidia dark, round, clavate, without ostiole, develop inside conidiophores of powdery mildew.
2. Conidia hyaline to sub-hyaline to dark, 1-celled, ovoid.
3. Hyperparasite on powdery mildew fungi.

***Coniothyrium* Sacc.**

1. Pycnidia globose, or flattened, black, conidiophores short, spores globose or ellipsoidal, small, brown, one celled.
2. Parasitic or saprophytic.

Sphaeropsis Sacc.

1. Pycnidia globose with ostiole, dark black.
2. Conidia one celled, oval to elongate large.

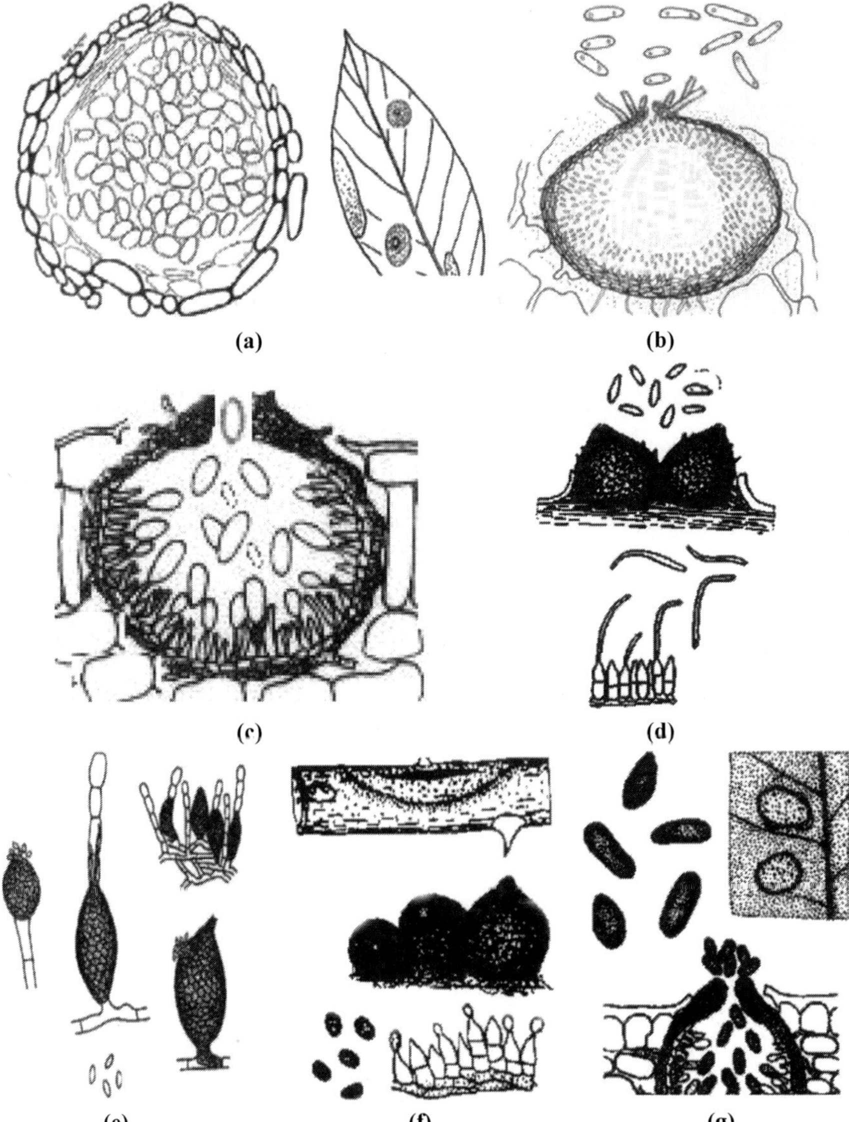

Fig. 6.43. Conidia and conidiophore of **(a)** *Phyllosticta* **(b)** *Phoma* **(c)** *Macrophoma* **(d)** *Phomopsis* **(e)** *Ampelomyces* **(f)** *Coniothyrium* **(g)** *Sphaeropsis*

***Ascochyta* Lib.**

1. Pycnidia, brown, thick-walled, immersed in leaf spots.
2. Conidogenous cells are annellides, enteroblastic, doliform or lageniform.
3. Conidia 2-celled, obovoid, hyaline to pale brown.

***Darluca* Cast.**

1. Pycnidia, brown, spherical, ostiolate, superficial on rust sori.
2. Conidia hyaline, 2-celled, ellipsoid to fusoid with bristle appendages at both ends.
3. Parasitic on rust fungi.

***Diplodia* Fr.**

1. Pycnidia-solitary, black, globose.
2. Conidia 2-celled, brown, ovoid or ellipsoidal
3. Parasitic or saprophytic.

***Botryodiplodia* Sacc.**

1. Pycnidia in clusters, thick walled, dark brown.
2. Ostiolate conidia-ellipsoidal or ovaid, 2-celled, dark brown striations at maturity.
3. Parasitic or saprobic on twigs.

***Septoria* Sacc.**

1. Pycnidia, dark, septate, globose, immersed, papillate in leaf spots.
2. Conidia-hyaline, filiform and multiseptate,
3. Causing leaf spots.

***Ephelis* Fr.**

1. Stroma resembles with smut gall, pycnidia erumpent, open, gelatinous.
2. Conidia hyaline, 1-celled, parasitic on grasses, conidial stage of *Balansia.*

***Hendersonia* Sacc.**

1. Pycnidia dark, separate, globose, ostiolate, immersed Conidia dark multi-celled, elongate to fusoid.
2. Parasitic or saprobic.

***Rhizoctonia* de Condole**

1. Mycelia sterilia, sclerotia without definite form, often grow together, horny- fleshy, with thinner undifferentiated edges, or embedded in mycelium.
2. *R. solani,R. bataticola* common in soil.

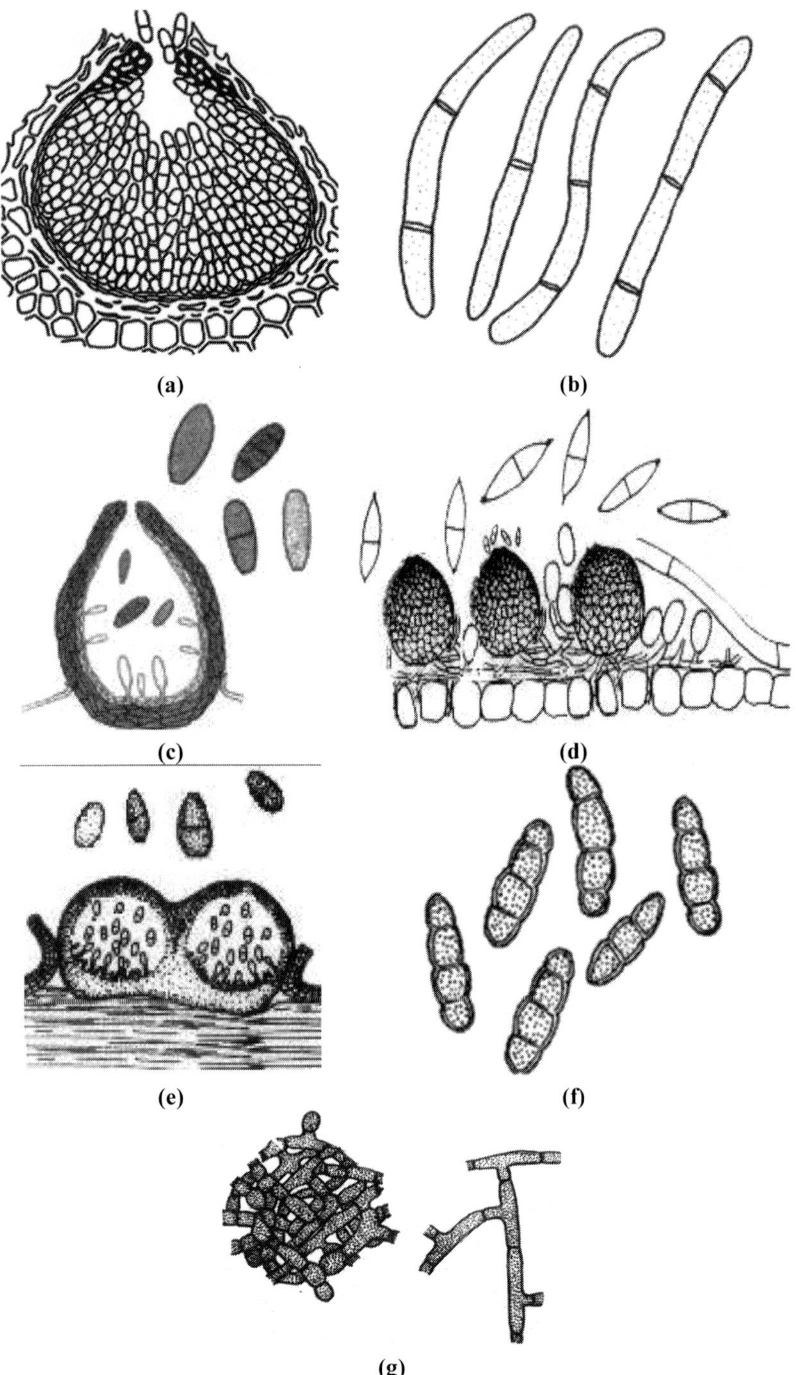

Fig. 6.44. Conidia and conidiophore of **(a)** *Ascochyta* **(b)** *Darluca* **(c)** *Diplodia* **(d)** *Botryodiplodia* **(e)** *Septoria* **(f)** *Hendersonia* **(g)** *Rhizoctonia*

Sclerotium Tode

1. Sclerotia variously formed, globose elongate, swollen or flattened, dark coloured, hard.
2. *S.rolfsii* is common.

Papulospora Preuss

1. Asexual spores absent, mycelium light to dark brown, produce compact clusters of bulbis (sclerotia).
2. Saprobic or parasitic.

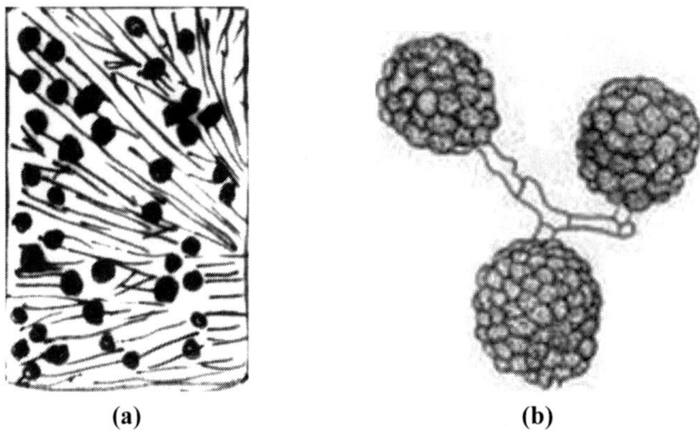

(a) (b)

Fig. 6.45. (a) *Sclerotium* **(d)** *Papulospora*

6.6.2 Class: Blastomycetes

Histoplasma

1. *H.capsulatum*, (the imperfect stage of *Ajellomyces capsulata.*).
2. Presence of large round of pear-shaped macrocondia with finger-like protuberances.
3. The microconida are small on short conidiophores, right angle to the hypha.
4. The fungus causes histoplasmosis to man, is an intracellular mycosis of lymphatic tissues, lungs, spleen, liver etc.

Sporothrix Hektoen and Perk.

1. *S.schenkii* responsible to cause sporotrichosis in man and animals.
2. This is chronic, sub-cutaneous mycosis affecting bones, lungs and central nervous system.
3. Presence of hyaline amerospores on short denticles, acrogenously on a simple sporogenous cell.
4. Dimorphic fungus with mycelia and conidial stages at room temperature and yeast phase at 37°C.

5. Two kinds of conidia, dark thin walled and hyaline.

***Geotrichum* Link.**

1. The vegetative hyphae fragments into unicellular units as arthrospores or arthroconidia.
2. *G.candidum*, imperfect stage of *Endomyces geotrichum.* In soil, dairy products and sewage.
3. Some isolates are pathogenic to man (cause geotrichosis) and other pathogenic to plants like lemon, tomato, and carrot.
4. Powdery nature of spores.
5. It grows vigorously on agar media, hyphae with broad septa, conidia holoarthric.

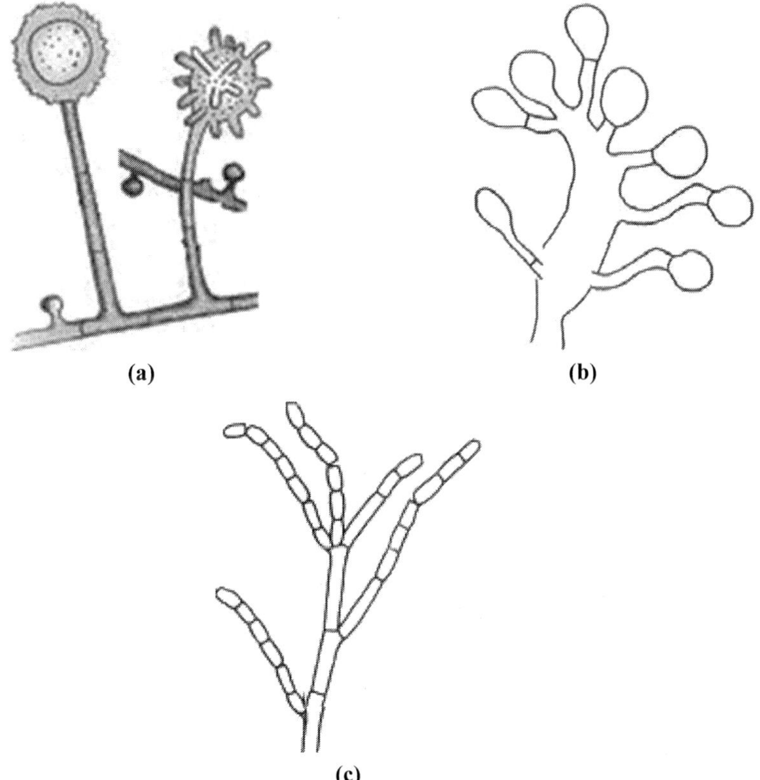

Fig. 6.46. (a) *Histoplasma* **(b)** *Sporothrix* **(c)** *Geotrichum*

***Trichophyton* Matmsten.**

1. Infect man and animals causing dermatophytosis (dermatomyces, tinea etc.), superficial fungus infection of keratinic tissues (epidermis, hair, and nails).
2. Conidia two types, the micro conidia smooth and thin-walled, hyaline, sub-spherical to clavate, borne singly or in grape like clusters.
3. Macroconidia smooth walled, spherical or clavate, blunt at the ends, many transverse septa.

Epidermophyton

1. Two species, *E.floccosum* and *E.cruris,* parasitic on man (glabrous skin) causing Tinea crusis, Tinea pedis, Tinea manun, Tinea corporis and onychomycosis.
2. Macroconidia paddle shaped, rounded ends, 4-5 cross walls, singly or in clusters on conidiophores.
3. Macroconidia absent
4. Chlamydospores abundant.

***Sporobolomyces* Kluyver and Van Niel.**

1. Grow saprophytically on ripe fruits, leaves, infected with smut or rust.
2. Vegetative phase yeast-like, uninucleated oval cells, secrete mucilage with pigments (red or pink), reproduction by budding.

***Cryptococcus* Kuts.**

1. In soil or decaying plant materials, few pathogenic to man and animals. Non–filamentous yeast, cell capsulated with ability to form starch.
2. The culture is mucous with red or orange colour.
3. The species are reported to be imperfect stages of Ustilaginales.

***Candida* Berkhout.**

1. Yeast-like organisms with well developed mycelial or pseudomycelial with multilateral budding
2. Present in soil, plant, and animal materials.
3. Some species pathogenic to man and animals.
4. *C. albicans* cause candidiasis (thrus) to pulmonary, volva etc.

***Rhodotorula* Harrison.**

1. Cell capsulated, do not produce starch, culture mucous.
2. Single celled yeast-like known as pink yeast.

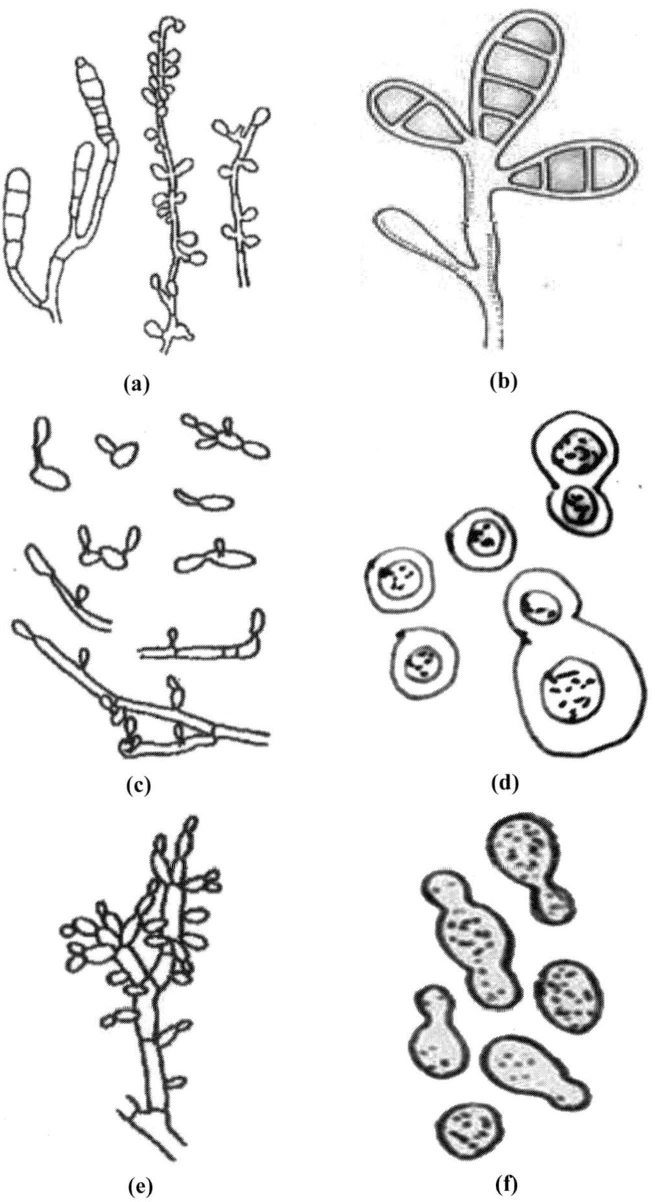

Fig. 6.47. **(a)** *Trichophyton* **(b)** *Epidermophyton* **(c)** *Sporobolomyces* **(d)** *Cryptococcus* **(e)** Candida krusei **(f)** *Rhodotorula*

7

Fungal Nutrition Culture Media

For growing fungi, a wide range of media are used. Media affect colony morphology and color. For utilizing different carbon sources some fungi lack the necessary enzymes. For growth and reproduction all fungi require several specific elements. The requirements for growth are generally less specific than for sporulation, so it is often necessary to try several types of media when attempting to identify a fungus in culture. Sporulating cultures should be used because repeated transfers of a pathogen on artificial media usually result in loss of sporulation or pathogenicity or both. Transfer should be made using spores and cultured alternately on nutrient rich and poor culture media. Most fungi thrive on potato dextrose agar (PDA), but in some fungi excessive mycelial growth is obtained at the expense of sporulation. Similarly, wood-inhabiting fungi and dematiaceous (dark pigmented) fungi often sporulate better on Corn meal agar (CMA) or Oat Agar, both of which have less easily digestible carbohydrate than PDA. Cellulose-destroying fungi grow on a weak medium such as Water Agar (WA) or Potato Carrot Agar (PCA). By adding pieces of tissue, such as filter paper, wheat straw, rice, grains, leaves or dung, often produces good sporulation dependent on the organism grown. Some special methods of inducing sporulation in fungi.

A. Seeding Culture Media

The type of inoculum and seeding method has quantitative and qualitative effects on the sporulation of most fungi. Seeding the medium with mycelia inoculum at one-point results in a small spore quantity, delayed sporulation, and spores of different ages. The medium should be seeded at various points using spores. A concentrated spore suspension spread over to yield the maximum spore number. The rapidity of sporulation and quantity of spores are related directly to spore concentration of the seeding suspension. If spores are not available, the medium should be seeded with a concentrated mycelial suspension, in which the mycelium may or may not be macerated.

B. Filter Paper Method

A method developed to induce sporulation in *Drechslera poae* and *Alternaria solani* has been used for inducing the sporulation in *Drechslera* state of *Cochliobolus sativus* and *Pyrenophora dictyoides*.

C. Grass-Leaf Method

This method is used to induce sporulation in *Alternaria*, *Colletotrichum*, *Curvularia*, *Drechslera*, *Fusarium*, *Helminthosporium* and *Nigrospora*. Small pieces of *Pennsetumum glaucum* leaves are autoclaved in water, the excess moisture removed by blotting on sterile blotter paper, and then transferred to culture plates with 1.5% water agar. The leaf strips or pieces are seeded with mycelia discs (previously described) obtained from 48 to 72 hrs. old colonies and incubated under optimum conditions for the test fungus.

Chopped leaves, stems, or roots are air dried, sterilized in a propylene oxide atmosphere for 24 to 48 hr, then plated on sterilized 1.5% water agar. The test fungus is placed at various points on the plant tissues and incubated at conditions optimum for growth.

The technique was used to induce sporulation in *Alternaria solani*, *Aschochyta pinodes*, and *A. pinodella* on pea straw, *Mycosphaerella*, *A. pinodes* on wheat straw, and *Phytophthora cinnamomi* oospores on avocado roots. Green leaves dried in a plant press and stored for future use or allowed to wilt overnight for immediate use can be used.

D. Cultivation of Pathogens for Soil Infestation

Large quantities of inoculum of soil borne pathogens can be obtained by using one of the following substrates in large containers.

1. Maize meal: Sand mixture (2-5:98-95), moistened and autoclaved for 1 to 2 h. on two successive days;
2. Vermiculite-mix (3:1) of the desired nutrient broth such as V-8: juice broth or potato-dextrose broth autoclaved for 1 to 2 h.
3. Grains of barley, maize, oat, pea, rice, or wheat individually or mixed are excellent substrates for many fungi; the grains are soaked in water for 24hr, drained and autoclaved.
4. A grain mixture with soil, sand, or vermiculite (1:1) can be used. Large flasks, prescription bottles, metallic or wooden trays can be used for containers.

Culture media and their compositions

Actinomycetes:

1. **Soybean meal-glucose agar**

   To 1 liter of water add 40 g soybean meal, 40 g glucose, and 3 g $CaCO_2$. Autoclave and filter the suspension. Dilute it to 8 liters, to each liter of the broth add 17 g Agar-agar and reautoclave. Adjust the pH 7.9 to 8.1.

2. **Starch (or glycerol) –casein agar**

Soluble starch (or glycerol)	10g	$M_gSO_4.7H_2O$	0.05g
$CaCO_3$	0.02g		
Casein (vitamin-free)	0.3g	$FeSO_4.7H_2O$	0.02g
KNO_3	2g	Agar agar	18g
NaCl	2g	Water	1000 ml
KH_2PO_4	2g	pH	7.0 to 7.2.

3. **Dextrose-nitrate agar**

Dextrose	1 g	$MgSO_4.7H_2O$	0.1 g
KH_2PO_4	0.14g	Water	1 000ml
$NaNO_3$	0.1g	Agar agar	15 g
KCI	0.1 g	Adjust to pH	7.0

4. **Egg albumin agar:**

Dextrose	1g	Egg albumin	0.25g
K_2HPO_4	0.5g	Agar-agar	15 g
$MgSO_4.7H_2O$	0.2g	Water	1000ml
$Fe_2(SO_4)_3$	trace	Adjust to pH	

Dissolve egg albumin in water make it alkaline using 0.1 N NaOH. Add other ingredients and adjust to pH 6.8 or near neutral before pouring into plates. For isolation of actinomycetes, add 400 mg cycloheximide just before pouring into plate.

5. **Glucose-asparagine agar:**

Glucose	10g	Agar-agar	15g
Asparagine	0.5g	Water	1000ml
K_2HPO_4	0.5g		

Adjust pH to near neutrality

6. **Glycerol-arginine agar:**

Glycerol	20g	$FeSO_4.7H_2O$	0.1g
L-arginine	2.5g	$MgSO_4.7H_2O$	0.1g
NaCl	1g	Agar agar	20g
$CaCO_3$	0.1g	Water	1000ml

Alternaria:

A. cuscutacide

7. **Maize (corn) seed-sunflower oil medium**

Maize (corn) seed	20g	$Ca(OH)_2$	0.5g

Sunflower oil	1ml	Sucrose	20g

Aphanomyces

A. raphani:

8. **Radish agar**

   Soak 200 g radish root tissue in 1000ml of water and keep at 60 °C for 1 h or use 250 g root tissue and keep for 45 min at 100 °C. Filter through cheese cloth and add 20 g Agar-agar agar to the filtrate.

9. **Radish-glucose agar**

   To the medium no. 8 add 10% glucose.

10. **Radish extract-peptone broth**

    To the medium no. 8 add 0.5% peptone and remove Agar agar.

***Aspergillus* spp**

11. **Sodium chloride-glucose-dicloran medium**

Peptone	5g	Glucose	10g
KH_2PO_4	1g	$MgSO_4.7H_2O$	0.5g
NaCl	30g	Agar agar	20g
Water	1000 ml		

After autoclaving and cooling to about 50^{0}C, add 50mg streptomycin, 50mg chlortetracycline, and 1mg dicloran (dissolved in 2 ml of acetone).

A. flavus

12. **Sucrose-peptone-dicloran agar:** Used to isolate and enumerate from soil.

K_2HPO_4	0.5g	$MgSO_4.7H_2O$	0.5g
Peptone	0.5g	Yeast extract	0.5g
Sucrose	20g	Rose Bengal	25mg
Streptomycin	50mg	Agar-agar	17g
Water	1000ml		

After autoclaving adjust to pH 5.5 using 1 NH4CI or NaOH. Add 5 to 10 mg dicloran pour into plates 3 to 4 days before use.

Ascomycetes

13. **Leonian agar:** Used to cultivate and sporulate

Peptone (or neopeptone)	0.625 g	KH_2PO_4	1.25g
Maltose (or glucose)	6.25 g	$MgSO_4.7H_2O$	0.625 g
Malt extracts	6.25 g	Agar-agar	20 g
Water	1000 ml		

14. **Botrytis sporulation medium**

Yeast extract	2.5g	Microelement stock solution (1 ml) contains/ lit	
Malt extract	7.5g		
Casein hydrolysate	0.25g	Fe(NO3)$_3$.9H_2O	723.5 mg
Na nuclaeate	0.1g	Agar-agar	15g
Microelement stock solution	1 ml	$ZNSO_4$. 4H_2O	203 mg
Agar-agar solution	15 g	H_3 BO_3	2M mg
Czapek-Dox broth	1000ml	H_2MoO_3	2mg

Botrytis alli* and *B. cinerea

15. Selective medium used for distinguishing between *B. alli* and *B. cinerea*

KCL	1g	Casein hydrolysate	5g
KH_2PO_4	1.5g	Glycerol	5g
$NaNO_3$	3g		

Coelomycetes

16. **Palm leaf decoction-palm wine agar:**

Finely grind 300 g palm leaf and mixed with 500 ml distilled water, steam for 1 hour and strain through a double layer of cheese cloth. Mixed the filtrate with 500 ml water in which 12 g agar-agar has been dissolved. Adjust the volume to 1000 ml. Sterilize by steaming for 1 hour. Sporulation does not occur if the medium is sterilized under pressure. Add to the autoclaved medium 60 ml of filter-sterilized palm wine.

Cercospora arachidicola:

17. **Glucose-phenylalanine agar**

Glucose	3 g	KH_2PO_4	1 g
Phenylalanine	0.5 g	$MgSO_4$.7H_2O	0.5 g
$ZnSO_4$	0.2 mg	Biotin	5 mg
$FeSO_4$	0.2 mg	Agar-agar	20 g
$MnSO_4$	0.1 mg	Water	1000 ml
pH	6.0		

18. **Sugarbeet leaf-decoction agar:** Finely grind 300 g of freshly picked young leaves of sugarbeet and add to 1.0 lit of water containing 12 g melted agar. Boil the mixture for 5 minutes, strain through cheesecloth, and autoclave the filtrate for 15 min at 110 to 115 °C.

B. elaeidis:

19. **Peanut leaf-oatmeal agar:**

Blend 50 g peanut leaflets in 500 ml water for 10 to 15 sec. and filter through cheesecloth. Boil 15 g oatmeal in 500 ml water for 15 minutes and filter through cheesecloth, combine equal amounts of both decoctions and add 20 g Agar-agar and autoclave.

Cercospora nicotianae:

20. **Tobacco leaf-decoction agar**

Harvest yellow-green leaves from the lower quarter of greenhouse-grown tobacco plants at blooming stage. Grind 600 g leaves in 2000 ml of water and steam, without pressure, for 1 h. Strain through cheesecloth, adjust to 2000 ml and add 60g agar-agar and autoclave.

C. beticola:

21. **Sugarbeet molasses agar:**

To 1000ml of water add 150g sugarbeet molasses, 15g Agar -agar and autoclave.

Cercospora. apii:

22. **Celery-leaf muck soil**

Place a 2 cm deep layer of moist muck compost with sand in a flask. Autoclave and then place the celery leaflets on the soil surface and re-autoclave for 20 minutes.

C. canescens:

23. **Carrot leaf-Oatmeal agar:**

Carrot leaf juice	50 ml	Agar agar	20 g
Oatmeal	50 g	Water	950 ml

Boil oatmeal for 5 minutes in water and filter it. Add carrot leaf juice to filtrate and make the volume to 1 liter.

Cochliobolus heterostrophus

24. ***Maize leaf agar***

Seedling corn leaves	125 ml	Agar-agar	20g
Sucrose	30 g		

Autoclave 125 ml of chopped seedling corn leaves in 875 ml water. Strain through cheese cloth and adjust to 1 litre. Amend with 30 g sucrose, 20 g agar-agar and autoclave.

C. sativus:

25. **Sach's agar**

$CaNO_3$	1g	K_2HPO_4	0.25g
$MgSO_4 \cdot 7H_2O$	0.25g	$CaNO_3$	4g
$FeCL_3$	0.025 g	Agar agar	20g
Water	1000ml		

Cochliobolus carbonum* and *Pyrenophora graminicola

26. **Lactose-casein hydrolysate agar:** Used to induce sporulation

Lactose	37.5 g	Microelement stock solution	
$MgSO_4.7H_2O$	0.5 g	$ZnSO_4.7H_2O$	439.8 mg
Casein hydrolysate	3 g	$MnSO_4.4H_2O$	203 mg
Microelement stock solution	2 ml		
$Fe(NO_3)_3.9H_2O$	723.5 mg		
Agar-agar	10 g		
Water	1000ml		

For preparation of microelement stock solution dissolve each salt one by one in 1 liter of water. To obtain clear solution add, drop by drop H_2SO_4 while solution is stirred. Adjust the pH to 6.0.

***Colletotrichum lagenarium*:**

27. **Green bean agar**

Finely grind 400 g of cooked green beans. Make up the volume to 1 liter with water and add 20 g agar agar and autoclave.

C. coccodes:

28. **SST medium**

Mix equal parts (w/w) of 2mm quartz sand, garden loam soil, and finely ground tomato roots from mature plants. Autoclave for 45 minutes. After 2 days adjust to 80% moisture holding capacity and re-autoclave for 15 min.

Soil extract-polygalacturonic acid agar

K_2HPO_4	4g	KH_2PO_4	1.5g
Soil extract	25ml	Polygalacturonic acid	10 g
Agar agar	17g	Water	1000ml

C. lindemuthianum:

29. **Bean juice agar:** Used to induce sporulation

Green beans juice	430 ml	Streptomycin sulfate	0.1 g
2 % molten agar	570 ml	Tetracycline	100 mg
Quintozene	100 mg	HCL	100 ml
Benomyl	0.1 g	chloramphenicol	100 mg

30. **Neopeptone-glucose agar:** Used to cultivate and sporulate

Glucose	2.8 g	Neopeptone	2g
$MgSO_4.7H_2O$	1.23 g	KH_2PO_4	2.72 g
Agar-agar	20 g	Water	1000ml

Glucose can be replaced by sucrose, xylose or galactose. Adjust to pH 5.2 to 6.5

Rhizoctonia bataticola:

31. **Flentze's soil extract agar:** Used to induce sporulation

Sucrose	1 g	Soil extract	1000ml
KH_2PO_4	0.2 g	Agar-agar	25 g
Dried yeast	0.1 g		

Mix 1 kg soil with 1000ml of water and keep for 1 to 2 days, agitate frequently. Filter through glass wool and make the volume to 1000 ml.

Cylindrocladium spp.:

32. **Peptone-dextrose agar:**

Dextrose	20 g	Peptone	0.2 g
Agar-agar	20 g	Water	1000ml

After autoclaving adjust the pH at 4 using 20% lactic acid.

Dematiaceous fungi:

33. **Hay infusion agar:** Used to cultivate and sproulate fungi

Autoclave 50 g decomposing hay in 1 liter of water for 30-minute, filter and adjust the volume of the filtrate to 1000ml. Add 2 g K_2HPO_4 and 15 g agar agar. Reautoclave for 10 min.

***Drechslera* and *Helminthosporium* spp.**

34. **Sucrose-proline agar (For sporulation)**

Sucrose	6g	$MgSO_4 \cdot 7H_2O$	0.5g
Proline	2.7g	$FeSO_4$	10mg
K_2HPO_4	1.3g	$ZnSO_4$	2mg
$KH_2 PO_4$	1g	$MnCl_2$	1.6mg
KCI	0.5g	Agar-agar	20g
	Water	1000ml	

Drechslera* spp. and *Pyricularia oryzae:

35. **Rice polish agar (For induce sporulation)**

Mix 20 g rice polish with 500 ml water and steam for 15 minutes. Blend the suspension for several minutes and mix with 500 ml water containing 17 g agar-agar. Autoclave and pour into culture plates. While pouring, agitate the medium to prevent settling of rice polish.

Fungi

36. **V-8 Juice agar: (Can replace PDA)**

The common formula is 200 ml of v-8 juice, 3g $CaCO_3$ with desired amount of agar agar and diluted to 1liter with water with pH 7 to 7.5 which can be varied by changing the amount of juice or $CaCO_3$. It is opaque medium, can be partially clarified by centrifuging or filtration through single, then double layer of filter paper.

37. **Cellulose medium Immerse**

3g small strips of Whatman no. 1 filter paper or powder cellulose in 100 ml concentrated HCI and let stand for 3 hr at 25 to 27°C. Shake occasionally. Pour the mixture into an excess of water. After 2 to 3 hr pour off the supernatant, wash the residue free of acid. Then wash with ethanol and air dry. The powder is used at a rate of 1% in Czapek's salt solution. Place the powder in a mortar and soak in a small quantity of Czapek's salt solution. Mix it into a fine paste using a pestle. Make the volume with Czapek's salt solution. Mix well and add agar agar at the rate of 1.5%.

38. **Cellulose medium for cellulolytic fungi from soil.**

$(NH_4)_2SO_4$	0.5g	Yeast extract	0.5g
L-asparagine	0.5 g	Cellulose	10g
KH_2PO_4	1g	Agar-agar	20 g
Crystalline $MgSO_4$	0.2g	Water	1000ml
$CaCl_2$ 0.	1g		

Add 10g Whatman cellulose powder (chromatography) to 250 ml of water and blend after 72 hours. Add to this other components dissolved in 750 ml of water. Autoclave for 20 minute at 110°C and adjust pH 6.2.

Fungi sporulation:

39. **Peptone Rose Bengal agar (Isolate and enumerate soil fungi)**

Peptone	5 g	KH_2PO_4	1 g
Dextrose	10 g	$MgSO_4.7H_2O$	0.5 g
Agar-agar	20 g	Rose Bengal	30 mg
Water	1000ml		

After autoclaving add 30 to 100 mg streptomycin or 30 mg aureomycin. Rose Bengal can be substituted by technical grade phosfon (2, 4 dirchloro-benzy ltributyl phosfonium chloride) at the rate of 500 mg/l.

40. **Carrot agar:** A weak medium used to induce sporulation in many fungi. Soak 20 g sliced carrots for 1 hr in 1000ml of water and then boil for 5 min. filter and add 20 g agar-agar.

41. **Alphacel agar:** For fungi cultivation and sporulation.

Alphacel	20 g	Coconut milk	50ml
$MgSO_4.7H_2O$	1g	Agar-agar	12g
KH_2PO_4	1.5 g	Water	1000ml
$NaNO_3$	1g		

Coconut milk filtered through cheese cloth and autoclave and can be modified by adding 10 g tomato paste and 10 g oatmeal.

42. ***PDA-Rose Bengal*:** Used to isolate and enumerate fungi from soil: Amend autoclaved PDA with 30 to 50 µg/ml rose Bengal and 50 to 100 µg/ml streptomycin.

43. ***Fomes annosus***

Peptone	5 g	$MgSO_4.7H_2O$	0.25 g
KH_2PO_4	0.5g	Quintozene	190 mg
Streptomycin	100mg	Agar-agar	20 g
Ethanol	50 ml	Water	1000ml

44. **Czapek-Dox agar:**

$NaNO_3$	2g	KCl	0.5g
K_2HPO_4	1g	$FeSO_4$	0.01g
$MgSO_4.7H_2O$	0.5g	Sucrose	30g
Agar agar	20g	Water	1000 ml

45. **Dextrose-peptone-yeast extract agar:** Used to isolate fungi from soil.

Dextrose	0.5 g	NH_4NO_3	5 g
Peptone	1 g	K_2HPO_4	1 g
Yeast extract	2 g	$MgSO_4.7H_2O$	0.5 g
Sodium propionate	1 g	$FeCl_3$ $6H_2O$	trace
Oxgall	5 g	Agar-agar	20 g
Water	1000ml		

46. **Malt extract agar:**

Heat 20 g malt extract in 1000ml of water until dissolved, then add 20 g agar agar and heat until the agar melts. The medium is between pH 3.0 to 4.0 and may be adjusted to 6.5 with NaOH. For cultivation of Basidiomycetes, use 50 g malt extract and 5 g malic acid.

47. **Malt and yeast extract agar:**

Malt extract	3 g	Yeast extract	2 g
KH_2PO_4	0.5 g	$MgSO_4.7H_2O$	0.5 g
Agar agar	20 g	Water	1000 ml

***Fusarium oxysporum* f. sp. *lycopersici*:**

48. **Cerclese-nitrate medium**

Cerelose	50g	NH_4NO_3	10g
KH_2PO_4	5g	$M_gSO_4.7H_2O$	2.5g
$FeCL_3O6H_2O$	0.02g	Water	1000ml

Fusarium graminearum.

49. **CMC-yeast extract medium:** Used to cultivate and induce sporulation

Carboxyl methyl cellulose	15g	NH_4NO_3	1g
Yeast extract	1g	KH_2PO_4	1g
$MgSO_4 7H_2O$	0.5g	Water	1000ml

***Fusarium oxsporum* f. sp. *melonis*:**

50. **Wensley and McKeen's medium**

Peptone	15g	$MgSO_{4.}7H_2O$	0.5g
KH_2PO_4	1g	Agar-agar	25g
Rose Bengal	35mg	Oxgall	0.5g
Water	1000ml		

Fusarium oxysporum* f. sp. *pisi.

51. **Cerelose-nitrate medium:**

Cerelose	30g	KNO_3	1g
$M_gSO_{4.}7H_2O$	0.5g	KCI	0.5g
F_eSO_4	trace amount	Water	1000ml

Fusarium oxysporum* f. sp. *lycopersici

Fusarium graminearum and Kabatiella zeae.

52. **Casamino acid medium:**

Glucose	15g	Casaminoacids	1.5g
Yeast extract	1g	KH_2PO_4	1.5g
Water	1000ml	Hoagland's solution	10ml

Fusarium spp:

53. **Potato-sucrose agar:**

Boil 200 g of peeled and sliced potatoes in 2 liter of water until the potatoes are soft, strain though cheese cloth and adjust the filtrate to 1 liter, add 20 g sucrose and 20 g agar agar, adjust pH. 6.5 with $CaCO_3$, autoclave.

54. **Malachite-captan agar:**

$NaNO_3$	2 g	K_2HPO_4	1 g
$MgSO_4.7H_2O$	0.5g	KCI	0.5 g
$FeSO_4$	0.01 g	Sucrose	30 g
Agar-agar	20 g	Water	1000ml

After autoclaving amend the cool, molten medium with 50 mg malachite green, 50 mg streptomycin sulfate, 50 mg captan, 75 mg dicrystine, 300,000 units of porcain penicillin, and 100,000 units of Na penicillin G.

55. **Joffe's medium used to induce sporulation in many *Fusarium* spp.**

KH_2PO_4	1 g	KNO_3	1 g
$MgSO_4$ $7H_2O$	0.5 g	KCI	0.5 g
Starch powder	0.2 g	Glucose	0.2 g
Sucrose	0.2 g	Water	1000ml
Agar-agar	15 g		

After autoclaving pour the medium into plates. Place sterile strips of cellulose lens paper over the agar surface before it solidifies.

56. **Kerr's medium:** Used to isolate *Fusarium spp.* from soil. Without amendments it is used to cultivate *Fusarium solani.*

$NaNO_3$	2 g	KH_2PO_4	1 g
KCI	0.5 g	$MgSO_4.7H_2O$	0.5 g
$FeSO_4$	0.01 g	Sucrose	30 g
Yeast extract	0.5 g	Water	1000ml
Agar-agar	15 g		

After autoclaving, amend the cool, molten basal medium with 60 mg rose bengal, 100 mg quintozene and 50 mg streptomycin.

57. **Komada's medium:**

L-asparagine	2 g	D galatose	20 g
$MgSO_4.7H_2O$	0.5 g	K_2HPO_4	1 g
KCI	0.5 g	Fe(EDTA)	5 mg
Water	1000ml	Agar agar	20 g

After autoclaving and cooling (45-50 °C) add 1 g glintozene, 05 g oxyall, 1 g NA_2 B_2 $O7.10H_2O$ and 30 mg streptomycin sulfate. Adjust to Ph 3.8 to 4.0 with phosphoric acid.

58. **Quintozene-peptone agar:**

Peptone	1.5 g	KH_2PO_4	1 g
$MgSO_4.7H_2O$	0.5 g	Agar agar	20 g
Streptomycin	300 mg	Water	1000ml
Quintozene	1g		

The medium need not be autoclaved. The culture plates, dried at 100 °C need not be sterilized.

***Fusarium solani* f. sp. *cucurbitae*.:**

59. **Gluose-isoleucine agar**: Used to induce perfect state

Glucose	10 g	D-L isoleucine	5 g
KH_2PO_4	1.75 g	$MgSO_4\ 7H_2O$	0.75 g
Agar-agar	20 g	Water	1000ml

Adjust the pH to 5.5 before autoclaving

Fusarium oxysporum f. sp. batatas

60. **Glucose-Sweet Potato Agar:**

To 1000ml of water add 18 g Agar-agar, 10 g glucose and some fragents of sweet potato roots (enlarged), sweet potato vines, or potato tubers.

***Fusarium oxysporum* f. sp. *lycopersici*.**

61. ***Fusarium* medium:**

Sucrose or glucose	2%	$MgSO_4.7H_2O$	0.003 M
KCl	0.022 M	KH_2PO_4	0.008 M
$FeCl_3$, $MnSO_4$, Ca $(NO_3)_2$	0.0356 M	ZnSO4, each	0.2 µg/ml

***Fusarium oxysporum:* (from soil)**

62. **Galactose-nitrate agar:**

Galactose	10 g	$K_2S_2O_5$	0.3 g
$NaNO_3$	2 g	Agar-agar	15 g
KH_2PO_4	1 g	Water	1000ml
$MgSO_4.7H_2O$	0.5 g		

Autoclave for 20 min

Fusarium oxysporum* f. sp. *pisi

63. **Azide Rose Bengal Agar:**

Dehydrated trypticase

Soybroth	30g	Rose Bengal	30 mg
KH_2PO_4	1.5g	Agar-agar	15 g
NaN_3	25 mg	Water	1000ml

Adjust the pH 7.6 before autoclaving and add 100mg quintozene, 30 mg streptomycin sulphate, 100 mg chlorotetracycline, HCl and 100 mg neomycin after autoclaving.

Fusarium sulphuram chlamydospores:

64. **Mannitol medium:** Used to convert conidia of *Fusarium sulphureum* to chlamydospores.

D-mannitol	23.7 g	K_2HPO_4	0.12 g
KH_2PO_4	0.4 g	$MgSO_4.7H_2O$	0.12 g
$(NH_4)_2\ SO_4$	0.3 g	Water	1000ml

F. roseum avenaceum

65. **Carnation dextrose agar:**

Prepare hot water extract from 200 g chopped carnation tissue, bring the volume to 1000ml add, 20 g dextrose and 20g agar agar. Autoclave

Gloeosporium musarum

66. **Glucose-peptone agar:**

Glucose	10 g	Peptone	2 g
KH_2PO_4	0.5 mg	$MgSO_4.7H_2O$	0.5 g
Agar agar	15 g	Water	1000ml

Glomerella cingulata:

67. Carrot juice agar

Carrot Juice	125-750 ml	Streptomycin or penicillin	100mg
Agar-agar	15 g	Water	1000ml
pH	5.1 to 5.5		

Helminthosporium oryzae **and** *Trichoconis padwikii*

68. **Guaiacol Agar:**

Guaiacol (O-methoxyphenol)	0.125 g	Agar-agar	5 g
Streptomycin	0.5 g	Water	1000ml

69. ***Macrophomina* sporulaton medium.**

Glucose	20 g	Peptone	20 or 40 g
DL-asparagine	15 or 30 g	$MgSO_4.7H_2O$	0.5 g
KH_2PO_4	1 g	Agar agar	20 g
Water	1000ml		

Autoclave and pour the medium into culture plates. Ether extract from peanut meal or cooking oil (blend of peanut oil and sunflower oil "COVO") is used to induce sporulation.

Macrophomina phaseolina:

70. **Chloroneb – ceresin agar**

Heat 1000ml of water to boiling and add 10g polished rice. Boil for 5 min. Immediately filter through cheesecloth. To the filtrate add 20g agar agar and autoclave. Cool the basal medium to 50 to 55 °C and add 150 mg chloroneb, 0.25mg actual mercury in the from of ceresin wet, 40mg streptomycin sulfate and 60 mg penicillin G. Adjust to pH 6. Ingredients must be fresh and added in sequence given.

71. **MRA medium**

Amend the rice agar medium no. 24 with 300 mg chloroneb, 7 mg $HgCL_2$, 90 mg rose Bengal, 40mg streptomycin and 60 mg penicillin G. The medium was modified by increasing the concentration of chloroneb to 312g, $HgCL_2$ to 8.5 mg and rose Bengal to 112 mg. adjustment of pH was omitted with a natural range of pH 7.5 to Good results are obtained if chloroned and rose Bengal are dissolved first in sterile cool water and $HgCL_2$ in sterile hot water prior to adding to the cool, molten medium.

72. **Soybean Decoction-Sucrose Medium:**

Boil 100 g soybean seed in 1 lit. of tap water and strain through cheesecloth. Adjust the volume of the filtrate to 1 lit, add 20 g sucrose.

73. **Maize meal agar:**

Corn meal powder	40g
Agar-agar	15g
Water	1000ml

Nectria cosmariospora:

74. **Prune agar**

To 1000ml of water add 40 g mashed prunes and 20 g agar agar.

***Phytophthora* spp:**

75. **Hendrix's medium**

Glucose	5.4 g	$NaNO_3$	1.5 g
KH_2PO_4	1 g	$MgSO_4.7H_2O$	0.5 g
Agar-agar	17 g	Thiamine HCI	2 mg
Water	1000ml		

Adjust to pH 6

76. **Rape-seed agar**

Boil 100 g rape seed in 1000ml of water for about 1 h. Filter the decoction. Adjust the volume of the filtrate to 1000ml and add 20 g agar agar.

Autoclave for 10 min. and pour 25 ml in each culture plate. After solidification spread 20 mg of sterol, dissolved in 2 ml ether over the agar surface. Use after several hours.

77. **Lima bean agar**

Blend 50 g lima beans, previously soaked in water for 10 hr with 500 ml of water containing 15 g melted agar. Adjust the final volume to 1000 ml and then autoclave.

78. **Lima bean decoction broth:**

Boil 100 g lima beans in 1 lit. of water, strain through cheesecloth, and complete the volume.

79. **MRA agar**

KH_2PO_4	30 mg	K_2HPO_4	30 mg
$MgSO_4.7H_2O$	20 mg	$CaCl_2$	0.56 mg
$MnCl_2$	2.68 mg	$ZnCl_2$	1.67 mg
$FeCl_2$	0.1 mg	Disodium salt of EDTA	11.06 mg
Sucrose	410 mg	L-asparagine	120 mg
Thiamine HCL	40 mg	Agar-agar	20g
Water	1000ml		

After autoclaving for 30 min, add to cool molten medium 5 mg each of endomycin and chloromycetin.

80. **Hendrix and Kuhlman's medium**

$NaNO_3$	2 g	$MgSO_4.7H_2O$	0.05 g
$FeSO_4$	0.01 g	KH_2PO_4	1 g
KCI	0.5 g	Sucrose	30 g
Yeast extract	0.5 g	Agar agar	15 g
Rose Bengal	60 mg	Water	1000ml

After autoclaving amend the basal medium (at 45 to 50 oC) with 100 mg quintozene, 100,000 units of mycostain, and 4.8 mg streptomycin sulfate. Adjust to pH 4.8 with lactic acid. The basal medium can be prepared in advance and stored in the dark at room temperature. At the time of use, it is melted, cooled, and amended.

81. **Klemmer and Lenny agar used to induce oospore formulation in *Phytophthora* spp.**

KH_2PO_4	680 mg	$MgSO_4.7H_2O$	250 mg
$Na_2HPO_4.7H_2O$	1.34 mg	$CaCl_2.2H_2O$	50 mg
$FeSO_4.7H_2O$	1 mg	$ZnSO_4.7H_2O$	4.4 mg
$CuSO_4.5H_2O$	0.08 mg	$NaMoO_4.2H_2O$	0.05 g
$MnCl_2.4H_2O$	0.07 mg	Thiamine HCI	100 mg
L-asparagine	708 mg	Glucose	15 g
Agar-agar	15 g	Water	1000ml

82. **Flower and Hendrix's medium**

$NaNO_3$	2g	$MgSO_4.7H_2O$	0.5 g
KH_2PO_4	1 g	Yeast extract	0.5 g
Sucrose	30 g	Thiamine HCI	2 mg
Gallic acid	425 mg		
Quintozene	25 mg	Rose Bengal	0.5 mg
Nystatin	100,000units	Penicillin G	80,000 units

83. **Glucose-asparagine-mineral salt medium**

KH_2PO_4	3 g	$MgSO_4.7H_2O$	0.5 g
$CaCl_2$	3.4 mg	$FeSO_4.7H_2O$	1 mg
$ZnSO_{4.}7H_2O$	1.8 mg	$MgSO_{4.}H_2O$	0.3 mg
$CuSO_4.5H_2O$	0.4 mg	$(NH_4)\ Mo_7O_{24}.4H_2O$	0.3 mg
Thiamine HCI	1 mg	L-asparagine	2 g
Glucose	20 g	Ergosterol	30 mg
Water	1000ml		

Prepare all minor elements and thiamine as 1000x stock solution and $FeSO_4.7H_2O$ in a citric acid solution at concentration of 1.4 mg/ml. Autoclave glucose separately. Dissolve ergosterol in 30 ml of dichloromethane and then add to KH_2PO_4 solution containing 0.5 ml Tween 80 and autoclave separately. Adjust the combined medium pH 6.0 with 1 N KOH.

84. **Maize (corn) Oil Agar:**

KH_2PO_4	1 g		
$CaSO_4\ 2H_2O$	0.1g	Dl threonine	1g
Thiamine HCl	0.02g	Sucrose	5g
Maize (corn) oil	0.1ml	Quintozene	0.1g
Triton B-1956	0.1ml	Agar-agar	20g
$MgSO_4.7H_2O$	0.5g	Water	1000ml

After autoclaving and cooling add 2mg Na pimaricin, 39,200 units nystatin dissolve in dimethys sulfoxide (DM SO) in to final concentration of 0.5% of DMSO in the medium, 60 mg chloramphenicol, 60 mg Na penicillin and 378,000 units of polymyxin sulphate.

85. **SV-8 M medium:**

To 1000ml of water add 10 ml of V-8 juice and 18 g agar. Autoclave and cool to about 50 °C: add 20 mg pimaricin, 75 mg vancomycin, 150 mg ampicillin, and 50 mg quintozene.

86. **Sucrose-asparagine-mineral salt medium:** Used for inducing oospore production

Sucrose	10g	Microelement stock solution contains per 100 ml:	
L-asparagine	1g	$Na_2B_4O_{7.}\ H_2O$	8.8mg
KH_2PO_4	0.5g	$CuSO_4 5.H_2O$	39.3
$MgSO_{4.}7H_2O$	0.25g	$Fe(SO_4)_3.9H_2O$	91 mg
Thiamine HCI	1 mg	$MnCl_2.4H_2\ O$	7.2 mg
Cacle anhydrous	0.1g		

Microelement stock solution 1 ml 1000ml

$NaMO_4.2H_2O$ 5g

$ZNSO_4.7N_20$ 440.3 mg

87. **Soybean meal agar:**

Grind 15 g soybean and add to 1 litre of water containing 20g agar.

$CaCl_2$(anhydrous)	0.1g	$MnCl_2 \cdot 4H_2O$	7.2 mg
microelement stock solution	1ml	$Na_2MoO_4 \cdot 2H_2O$	5mg
$ZnSO_4 \cdot 7H_2O$	440.3g	Water	1000ml
EDTA	50g		

88. **Mc-Intosh's agar (modified):**

$KH_2\ PO_4$	1g	$CaSO_4.7H_2O$	1g
DL-threonine	1g	Thiamine HCL	0.02g
Sucrose	20g	Mycostatin	10g
$MgSO_4.7H_2O$	1g	Agar-agar	20g
Na taurocholate	0.2g	Water	1000 ml

After autoclaving, add 20 mg benomyl dissolved in 5 ml of dimethyl sulfoxide.

89. **Maize (corn) meal-pimaricin-polymixin agar:**

Corn meal powder	40g	Polymyxin B Sulphate	370,000 units
Pimaricin	2mg	Agar-agar	15g
Quintozene	100mg	Water	1000ml
Penicillin G	80000 units		

Adjust pH 4.6

90. **Maize (corn) meal-pimaricin-vencomycin agar:**

Corn meal powder	40g	Agar-agar	15g
Pimaricin	5mg	Water	1000ml
Vencomycin	300mg	Adjust	pH 4.6

91. **Maize (corn) meal-pimaricin-vencomycin quintozene agar:**

Corn meal powder	40g		
Pimaricin	10mg	Agar-agar	15g
Vencomycin	200mg	Water	1000ml
Quintozene	100mg	Adjust pH	6.0

Phytophthora

92. **McIntosh's agar:** Used for *Phytophthora* spp. isolation

KH_2PO_4	1 g	$MgSO_4.7H_2O$	1 g
$CaSO_4.7H_2O$	1 g	Thiamine HCI	0.02 g
DL-threonine	1 g	Mycostatin	10 mg
Sucrose	20 g	Agar-agar	20 g
Water	1000ml		

Thiamine HCI, DL-threoconine, and mycostatin are added after autoclaving. Mycostatin is first dissolved in dimethyl sulfoxide (DMSO) and then added to the cool, molten medium to give concentration of 10 μg/ml of mycostatin and 0.5% of $DMSO_4$.

Pythium

93. **Chickpea agar:**

After washing 250g chickpea seed soak in distilled water overnight. Decant the water and mash the seed. Add 1lit. of water and steam for 1 h. Filter through cheesecloth. Dissolve 15g Agar-agar and 20g sucrose, separately, and then add to the filtrate. Make the volume to 1lit. autoclave for 15 nub at 110°c.

94. **Lima bean-dextrose agar:**

Very finely ground dry Lima bean	15 g	Dextrose	10 g
Yeast extract	2 g	Water	1000ml
Agar agar	10 g		

Phytophthora megasperma f. sp. glycinea:

95. Erwin and Mc Cormic's medium

Clarified V-8 juice	200 ml	$CaCO_3$	3 g
B-sitosterol	30 mg	Agar-agar	15 g
Water	800 ml		

Phytophthora cinnamomi

96. **Pea seed broth:**

Blend 200 g frozen peas in 500 ml of deionized water. The mixture will be centrifuged for 10 min at 4000 RPM. Decant the supernatant and adjust its volume to 1000ml.

Phytophthora phaseoli:

97. **Oatmeal agar:**

Soak 50 g rolled oats for 10 hr in 500 ml water and then blend in a commercial blender. Mix with 500 ml of water containing 15 g melted agar and autoclave

98. **Hemp seed agar:**

Boil 100 g hemp seed in 1000ml of water, filter, adjust the volume to 1000ml, and add 20 g Agar-agar. For cultivation of *P. middletoni*, grind 20 g seed in 1000 ml of water, filter through muslin cloth, and remove the seed coats. Add 20 g Agar-agar and autoclave.

99. **Vaartaza's medium:**

Sucrose	2g	Yeast extract	0.2g
K_2HPO_4	0.7g	KNO_3	1g
$CaCO_3$	0.5g	Agar-agar	20g
Water	1l		

100. **Schmitthenner's medium:**

Sucrose	2.5g	Asparagine	270 mg
KH_2PO_4	150mg	K_2HPO_4	150mg
$MgSO_4.7H_2O$	100 mg	$CaCl_2$	55mg
Thiamine HCI	2mg	Cholesterol	10mg
$ZnSO_4.7H_2O$	4.4 mg	$FeSO_4.7H_2O$	1g
$MnCl_{2.}4H_20$	0.07mg	Water	1000ml

101. **Hemp seed-carrot extract agar:** Used to cultivate and induce oospore production in *Pythium spp.*: Boil 20 g cracked hemp-seed in 1000 ml of tap water. Add 20 g agar agar and extract from 100 g carrots.

Pythium aphanidermatum:

102. **Maize (corn) meal-pimaricin-rose bengal-benomyl agar:**

Corn meal powder	40g	Benomyl	5mg
Pimaricin	100mg	Agar-agar	15g
Streptomycin	200mg	Water	1000ml
Rose bengal	150mg	Adjust pH	6.0

Rhizoctonia solani

103. **Pea seed medium:**

Soak dried yellow peas overnight in water. Place a 2- to 3- cm deep layer of pea in flask and add enough water to cover them and autoclave.

104. **Oatmeal agar:**

Blend, in a commercial blender, 75 g of rolled oats in 600 ml of water and then heat to 45 to 55°C, add 20 g agar agar dissolved in 400 ml of water. Autoclave for 90 min.

***Phytophthora infestans*:**

105. K_2HPO_4 1g Mg $K_2SO_4.7H_2O$ 0.5 g

K_2HPO_4	1g Mg	$K_2SO_4.7H_2O$	0.5 g
KCI	0.5 g	$FeSO_4.7H_2O$	10 mg
$NaNO_2$	0.2 g	Gallic acid	0.4 g
Agar agar	20 g	Water	1000 ml

After autoclaving, add to the cool molten medium 90 mg fenaminosulf, 50 mg chloramphenicol, and 50mg streptomycin.

106. **Potato-marmite agar** used to induce sporulation in *Rhizoctonia solani* and *Sclerotium rolfsii*

107. B-R medium:

Glucose	40 g	$MgSO_4.7H_2O$	200 mg
K_2HPO_4	696 mg	$FeSO_4.7H_2O$	10 mg
KCl	149 mg	$MnSO_4.H_2O$	6.2 mg
$ZnSO_4.7H_2O$	8.3 mg	Agar-agar	20 g
Thiamine HCl	0.1 mg	Water	1000 ml
NH_4NO_3	1g		

108. **Brown's medium:**

Glucose	2 g	$MgSO_47H_2O$	0.75g
K_2HPO_4	1.25 g	Water	1000ml
Asparagine	2 g		

109. **Potassium oxalate-gallic acid agar:** Used to selectively isolate *Sclerotium rolfsii* from soil.

KH_2PO_4	1 g	$MgSO_4.7H_2O$	0.5 g
KNO_3	2 g	Thiamine HCI	1 g
Galli acid	160 mg	Potassium oxalate	10 g
Microelement stocks solution	10 ml	Water	250 ml

Sclerotium cepivorum:

110. **Inulin-quintozene Agar**:

Inulin	12 g	$NaNO_3$	3 g
K_2HPO_4	1g	$MgSO_4.7H_2O$	0.5 g
Sodium ferric diethyl	1 g	Agar-agar	20 g
Netriamine pentacetate	5 mg	Water	1000 ml

Dissolve the inuline in hot water and cool to 40 °C, then add other ingredients. Adjust to pH 5.2 and autoclave for 30 min. after cooling; amend the basal medium with 1 g quintozene, 50 mg chlorotetracycline HCI, and 100 mg streptomycin sulfate.

***Septoria* spp.**

111. **TCV agar:**

In seeds: amend 1000ml of autoclaved PDA at 60°C with 5 ml solution containing 0.02% thiophanate methyl, 0.17% chloroneb, 0.02% vancomycin HCl, and four drops of Tween 20.

Septoria nodorum

112. **Sporulation agar:**

Carbon source	20 g	Nitrogen source	0.425 g
$MgSO_4.7H_2O$	0.5 g	$FeSO_4$	0.2 mg
$ZnSO_4$	0.2 mg	$MnSO_4$	0.1 mg
Agar-agar	20 g	Water	1000 ml

Septoria avenae:

113. **Oat leaf agar:**

Grind 300 g of fresh green oat leaves and mix the pulp and juice with 1000ml of water containing 12 to 15 g Agar-agar. Boil the mixture for 5 min, strain through cheesecloth, and autoclave the filtrate.

***Stemphylium* spp:**

114. Phytone-dextrose agar

Phyton	15 g	Dextrose	15 g
Yeast extract	1 g	Agar-agar	17 g
Water	1000 ml		

Verticillium:

115. Sorbose agar

Sorbose	2 g	Agar-agar	10 g
Streptomycin	0.1 g	Water	1000 ml

After autoclaving add to the cool, molten medium 50 mg cycloheximide and 50 ml of filter sterilized 10% solution of sorbitol.

Verticillium

116. **Glucose-mineral salt medium:**

Glucose	30 g	KNO_3	2 g
K_2HPO_4	1 g	K_2HPO_4	0.9 g
$MgSO_4.7H_2O$	0.5 g	$ZnSO_4.7H_2O$	0.05 mg
$MnSO_4.7H_2O$	0.5 mg	$CuSO_4.5H_2O$	0.16 mg
$NA_2MoO_4.2H_2O$	10 µg	Water	1000ml

8

Electron Microscopy

Diagnosis of plant diseases require some knowledge of light microscopy to examine infected tissues for the presence of plant pathogens. Most of the pathogens and the diseases caused by them can be identified quickly by using different methods such as section or cutting of infected tissue, squashes of scrapping of plant surface. Homecare, the detail study enquires ultrastructural examination of the pathogen.. Electron microscope rely on a beam of electrons produced from a hot filament (e.g. tungsten) and its power of resolution is much greater than light microscope. Electron microscope are of two type:-

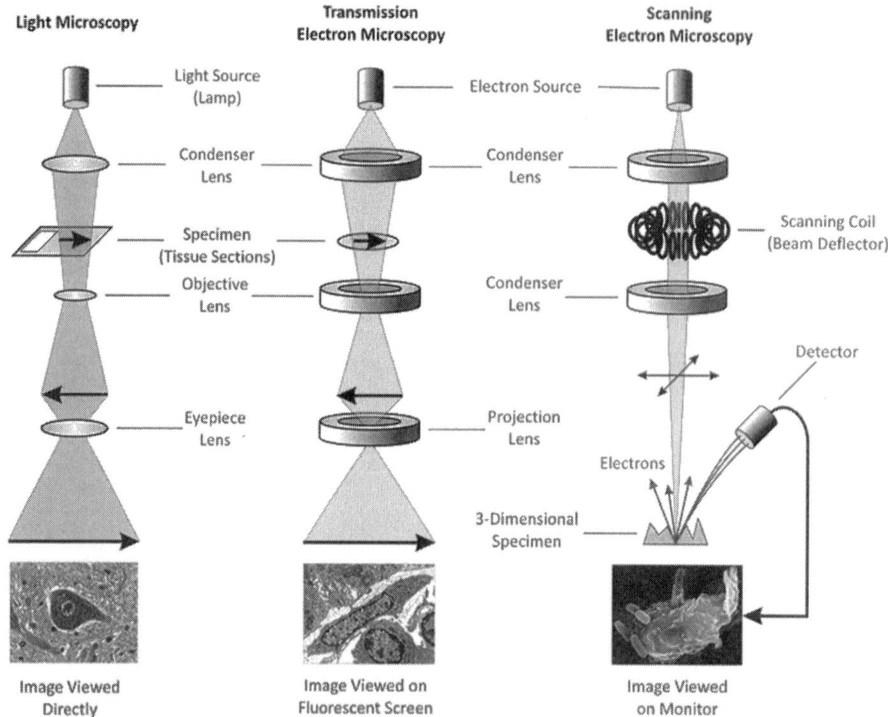

Fig. Comparative diagram of a light micrsoscopy, transmission and Scanning electron Microscopy

8.1 Transmission Electron microscope (TEM)

In transmission electron microscopy (tem) to form an image a beam of electrons is transmitted through a specimen.

The specimen is an ultrathin section less than 100 nm thick or a suspension on a grid. By the interaction of the electrons with the sample an image is formed. The image is then magnified and focused onto an imaging device, such as a fluorescent screen, a layer of photographic film.

Transmission electron microscopes are capable of imaging at a significantly higher resolution than light microscopes. This enables the instrument to capture fine details-even as small as a single column of atoms, which is thousands of times smaller than a resolvable object seen in a light microscope. Transmission electron microscopy is a major analytical method in the physical, chemical, and biological sciences. TEMs find application in cancer research, virology, and materials science as well as pollution, nanotechnology, and semiconductor research.

The first TEM was demonstrated by Max Knoll and Ernst Ruska in 1931, with this group developing the first TEM with resolution greater than that of light in 1933 and the first commercial TEM in 1939. In 1986, Ruska was awarded the Nobel Prize in physics for the development of transmission electron microscopy.

8.2 Scanning Electron microscope (SEM)

A beam of focused electrons is passed through a thin section of the specimen in TEM. Electron magnetic lenses magnify the image formed in the beam before the final image is projected onto a fluorescent screen where it can be photographed.

Fixed tissues and pathogenic fungi held in plate (plastic) and sliced over a hundred times thinner than a sheet of tissue paper to observe fine structure within a all this slicing is done within an ultramicrotome using an angular broken edge of electron beam on a plastic thin enough to be transparent to the electrons. This film is held by miniature wire mesh screen (grid) about 1/8 in diameter. This grid is held in a movable holder for observation in TEM.

The beam of focused electrons in SEM is scanned across the specimen by a beam of electron, which strikes a specimen emitting secondary electrons, which are attracted towards a passively charged grill and detection system.

These generate variable strength electronic signals, which offer amplifications and progressing, produce a final image in the cathode ray tube screen that can photographed.

8.2.1 Procedure

8.2.1.1 Fixation: since only defaces layers of the specimen are usually viewed with the SEM, the fixative used should be isotonic so as not to expose the cells to osmotic shocks Gletaradehyde (5%) in 0.5% phosphate buffer, (P^H7.2) is commonly used fixative for SEM studies.

8.2.1.2 Immersion: Excise the desired tissue and submerge in fixation medium

8.2.1.3 Dripping: This approach involves expanding the tissue surface to be studied by partial dissecting of a tissue and flooding it with large volume of the fixative into tissue.

8.2.1.4 Post fixation treatment: Expose the specimens to vapors of osmium tetra oxide (OS_2 O_4) prepared in phosphate buffer (P^H 6.8) for z to 24 hrs, depending upon the plant material.

8.2.1.5 Dehydration: In SEM, there is a need to remove the water from the Specimen. For dehydration, these processes can be used:

i. **Vacuum drying:** water is evaporated by exposing the tissue to vacuum drying.
ii. **Critical point drying:** The purpose of this process is to remove water from specimen, without damaging tissue by the surface tension which act to collapse structure as the receding liquid evaporates. There is two phase system, liquid water of the specimen and gaseous water over it. These is a critical point, in term of environmental temperature and pressure applied to liquid, where they become indistinguishable at this phase, can be converted into the gas phase. Ethanol is used as dehydrating agent and pressurized liquid CO_2 and at room temperature as the transitional liquid. The final step takes place in drying bomb, that preferably has a transparent window, so that the exchange can be followed visually. The temperature of the sealed bomb containing liquid CO_2 is raised above 31°C; the critical point of CO_2 to leave the bomb, the specimen tries.
iii. **Freeze Drying:** The technique involves the total removal of the specimen water of solvent by freezing it and submitting it off in a high vacuum system. In this technique the specimen is first fixed and dehydrated with ethanol and then substituted with amyl acetate for the ethanol prior to freeze drying is carried out at –70°C in vacuum of 10^{-1} to 10^{-3} fore. When freeze drying is completed, the specimen is warmed to a little above a room temperature to parent sarfea condensation of water prior breaking the vacuum.

iv. **Specimen mounting:** Large specimens are best mounted after drying. They may glued *e.g.*, with conductive coating paint. Apiezon wax or epoxy resin or attached by other means *e.g.*, double sealed adhesive tape to a standard SEM specimen holder.

v. **Specimen coating:** A specimen that is conductor when examined in SEM can be maintained at a desired patention since any change imposed on it by the incident electron beam can be led off. However, on insulated specimen exposed to the electron beam will become charged in a beam. This obstacles can be addressed by vacuum coating the specimen surface with a very thin layer of conducting material (0.01 to 0.1 mm thick), so as not to obscure the specimen surface, Electrical connection is made between the film and the specimen stage, so that charge is carried away.

Coating method

i. **Vacuum evaporation:** In evaporation until where the pressure of its vacuum chamber is reduced to 10⁻5 torr, the heated coating material rapidly evaporates into monatomic state. The vapour molecules are essentially unimpeded by are molecules luaus the source and move in a straight line until they impinge on the surface of the specimen, where they from a thin filn.

ii. **Sputter coating:** Sputtering is progress that occurs when an appropriate target metal is bombarded by energetic particles. A "gas" of metallic atoms is created in the poor vacuum by prolonged erosion of the target surface and they deposit uniformly as a coat on all specimen surfaces. Regardless of orientation.

Coating material

i. **Metal coating material:** A wide variety of metal have been used, along or in combination with gold, gold palladium or palladium are the most popular metals.

ii. **Carbon coating:** Use of carbon is advantageous because it easily evaporates, causes little specimen heating, and easily scattered to form a continuous layer to form tough substitute which stabilize the specimen surface.

9

Preservation of Fungi

When any serious work is undertaken involving the use of living cultures it is necessary to keep these alive for use during the work and for future reference.

This can be done either by growing on a suitable substrate and transferring to fresh media, as the nutrient is used up or by treating the cultures in some way, which halts metabolism until they are revived for use. The former method is time consuming and cultures are at risk whenever they are handled, so various means are devised which can lengthen the periods between sub culturing. The latter methods tend to be complex and expensive.

Methods of culture maintenance

1. **Frequent Transfer**

   Cultures are transferred from one agar slant or other suitable medium as required using the most satisfactory conditions. If screw cap bottles are used, caps must be kept loose to allow ventilation, but the bottles are sealed for protection against mites, using cigarette paper.

   Cultures are usually grown in the daylight at room temperature, but other conditions of light and temperature are used where necessary. Glass containers are standard but plastic ones are used if black light is necessary to obtain sporulation.

   These cultures are stored at room temperature, for short term storage in the refrigerator at 5-8°C. Transfer at 4-6 months interval. Some working collections are kept in this way.

   In the deep freeze survival is expected to be about 4-5 years. The method is suitable for all cultures, but it has certain disadvantages as a preservation technique.

   i. Danger of variation, either a change in desirable physiological responses or loss of morphological characters.

   ii. Contamination by aerial borne spores or mite carried infection.

   As a long term means of preservation, it requires continuous specialist supervision.

2. Storage under Mineral oil

Healthy cultures are covered with mineral oil, which appears to lengthen their life. The oil allows slow diffusion of gases and so growth continues at a reduced rate. However, this also encourages adaption to growth in oil.

i. Cultures are grown on suitable agar slants to give maximum sporulation or optimum growth.

ii. The slants are covered by mineral oil (British Pharmacopoeia quality, specific gravity 0.830-0.890 g commonly known as liquid paraffin or medicinal paraffin oil) to 1 cm above the top of the culture, to prevent drying out.

Retrieval is by removal of a small piece of mycelia, draining off as much of the oil as possible and streaking out on an agar plate.

The oil is double sterilized at 151b/in^2 for 15 minutes and is added as individual doses to prevent blow up and cross contamination. Care must be taken when sterilizing oily needles as they tend to splutter passing fungus into the atmosphere. This can be dangerous if handling pathogens. Rubber liners are removed from bottle caps as they can become dissolved in the oil, producing materials toxic to the cultures. The oiled cultures may be stored at 15°C, through. They can be kept at room temperature. Sub culturing of most lengthens life, many isolated being viable after 5, 10, or even 20 years. However, very sensitive isolates (*e.g.*, the chytrids) survive only 12 to 18 months. It is helpful for maintaining these, as they normally have to be subcultured every month and this period can be lengthened to a six month collections with limited resources, and in the tropics, as mites are repelled by the oil. Many fungi such as *Phytophthora* and *Pythium* species, which will not survive freeze drying, they do well in oil. Strain stability is not reliable, and some cultures deteriorate rapidly, *e.g. Fusarium* sp. and ascosporic *Penicillia.*

3. Silica gel storage

Spores stored in silica gel especially when protected by skimmed milk, remain viable for long periods, *e.g.*, 4-5 years. As heat is liberated when water is added to silica gel, the spore suspension and gel should be precooled before starting the process.

Small bottles, the size is immaterial, with screw caps, are partially filled with silica gel (purified silica gel, without indicator, 6–22 mesh) sterilized with dry heat (180°C for 3 hours) then precooled in ice or a refrigerator. A spore suspension in 5% precooled skimmed milk is added (dropped from a sterile pipette) to wet ¾ of the silica gel, the bottles are left in the cooling bath until they cool down (about 20 minutes) and then stored with loose caps at room temperature for 1-2 weeks, until the crystals readily separate. The caps are then screwed down and the bottles stored over indicator silica gel in an airtight container at 4°C. The gel can also be stored at room temperature with very good revivals. Retrieval is by scattering a few crystals on to a suitable medium, standard inoculum can be obtained from one bottle. Good revivals are obtained from most sporing cultures. The method is frequently used by workers undertaking genetic studies as cultures stored in this way appear particularly stable and were originally devised for this use (Ogata 1962). Silica gel storage is recommended to laboratories which wish to undertake a simple, stable, long term method of storage to whom freeze drying is not available.

4. **Soil storage**

One ml of a spore suspension in sterile water is added to 5.0 g of garden loam (20% moisture content, autoclaved twice for 15 minutes at 15 lb/in2) and allowed to grow at room temperature for 10 days. The culture are then stored in a refrigerator at 5-8°C, with loose caps. The longevity of culture is usually increases and morphological changes reduce. As cultures are obtained by the removal of a few grains of soil, constant, repeatable and stable inoculums is obtained over a period of time.

This method is especially used for *Fusarium* sp. which is renowned as unsuitable in culture and remains alive and typical by this method for a long period. The growth period at room temperature is usually reduced to 2-3 days in this instance.

5. **Water storage**

In 1976 Boesewinkel reported storing plant pathogenic fungi in sterile distilled water. This method has been used at the CMI to store some *Phytophthora* and *Pythium* species and has proved satisfactory over 2 years.

Small agar blocks about 60 mm3 are cut from the growing edge of the fungal colony and are transferred aseptically to small sterile screw cap bottles, half filled with sterile distilled water. The cap is screwed down and the bottles stored at room temperature. Retrieval is made by the transfer of one of the agar blocks to a suitable medium. The method is very simple and promises to be very useful and inexpensive form of storage.

Bibliography

Ainsworth, G.C. and Sussman, A.S. (1965). The Fungi: an advanced treatise Eds. Vol. I, "The Fungal Cell", Academic Press, New York.

Ainsworth, G.C. and Sussman, A.S. (1966). The Fungi: an advance treatise, Vol. II, "The fungal Organism", Academic Press, New York.

Ainsworth, G.C. and Sussman, A.S. (1968). The Fungi: an advanced treatise Eds. Vol. III, "A Taxonomic Review with Keys: The Fungal Population". Academic Press, New York.

Ainsworth, G.C.; Sparrow, F.K. and Sussman, A.S. (1973). The Fungi: an advance treatise, Vol. IVA, "Ascomycetes and fungi Imperfecti", Academic Press, New York.

Ainsworth, G.C.; Sparrow, F.K. and Sussman, A.S. (1973). The Fungi: an advance treatise, Vol. IV B, "Basidomycetes and Lower Fungi", Academic Press, New York.

Alexopoulos, C. J.; Mims, C. W. and Blackwell, M. (1996). "Introductory Mycology", 4th Eds., John Wiley and Sons Inc., New York.

Anton, S.M. Sonnenberg; Johan, J.P. Baars; Patrick, M. HendrickX; Brian, Lavrijssen; Wei, Gao; Amrah, Weijn snd Jurriaan J. Mes (2011). Breeding and strain protection in the button mushroom Agaricus bisporus, Proc. 7th Int. Conf. Mushroom Bio. Mushroom Prod., Arcachon, France.

Ashwathi, P. (2017). Reproduction in fungi http://www.biologydiscussion.com/fungi /reproduction-fungi/ reproduction-in-fungi-with-diagram-microbiology/49923.

Bessey, E.A. (1950). "Morphology and Taxonomy of Fungi". Blakiston Co., Philadelphia, USA.

Burnett, H.L. (1972). "Illustrated Genera of Imperfect Fungi", 3rd Eds., Burgess Publishing Co. Minneapolis.

Burnett, J.H. (1976). "Fundamentals of Mycology", Edward Arnold, London. Christensen C.M. (1965). The Molds and Man, 3rd Eds. University of Minnesota Press, Minneapolis.

Mehrotra, R. S. and Aneja, K.R. (1990). "An Introduction to Mycology", New Age International, Publisher, Hyderabad, Telangana.

Mukadam, D.S. (1997). "The Illustrated Kingdom of Fungi", Aksharganga Prakashan, Aurangabad, (M.S.).

Raper, J.R. (1954). Life cycles, sexuality, and sexual mechanisms in the fungi. (Wenrich, D.H., Lewis, I. F. and J.R.), American Association Advance in Sciences.

Raper, J.R. (1966a). Genetics of Sexuality in Higher Fungi. Ronald Press. New York. Raper, J.R. (1966b). Life cycles, basic patterns of sexuality and sexual mechanisms. In: The Fungi. Vol II. Eds. Ainsworth, G.C. and Sussman, A.S. Academic Press, New York, London. p. 473-511.

Russell, P. J. (2010). iGenetics: A Molecular Approach, 3rd Eds., San Francisco, CA: Pearson Education, Inc.

Sansome, E. (1966). Meiosis in the sex organs of the Oomycetes. In Chromosomes Today. Eds. C.D. Darlington and K.R. Lewis. Plenum Press, New York. p. 77-83.

Sarbhoy, A.K. (2000). "Textbook of Mycology", ICAR, New Delhi.

Singh, R.S. (1982). "Plant Pathogens: The Fungi". Oxford and IBH, Delhi.

Webster J. (1980). "Introduction to Fungi". 2nd Eds., Cambridge University, Press, Cambridge, New York.

Webster, J.W. (1970). "Introduction to Fungi", Cambridge University Press, London. Webster, J. and Weber, R. (2007). "Introduction to fungi", Cambridge University Press, 3rd Eds., New York.

Subject Index